VIA 7

THE BUILDING OF ARCHITECTURE

Editors

PAULA BEHRENS

ANTHONY FISHER

Production Editor

RICHARD MOHLER

The Graduate School of Fine Arts

University of Pennsylvania

and

The MIT Press

1984

Via *is the architectural journal of the Graduate School of Fine Arts, University of Pennsylvania. It is edited and designed by students of the Department of Architecture and is published annually.*

Staff

Andrew Blanda
James Dart
Ruth Finch
Erik Johnson
Chris Macneal
Robert Petito
Sally Pinigis
Richard Votta

Copy Editors

Deborah Gorman
Louis Gross

A. T. Seymour, III's article was previously published in *Texas Contractor* November 21, 1972.

An earlier article on the subject "Chinese Building Standards in the 12th Century", by Else Glahn was published in *Scientific American* May, 1981.

Parts of the article by Alexander Tzonis and Liane Lefaivre appeared in an earlier version in *Publication Series in Architecture*, Graduate School of Design, Harvard University, 1980.

Send orders and business correspondence to: Journals Department, The MIT Press, 28 Carleton Street, Cambridge, Massachusetts 02142.

Send editorial correspondence to: VIA, Room 102, Meyerson Hall/CJ, University of Pennsylvania, Philadelphia, Pennsylvania 19104.

In the United Kingdom, Continental Europe, the Middle East and Africa, send orders to: The MIT Press, Ltd., 126 Buckingham Palace Road, London SW1W9SD England.

Library of Congress: 77-76853
ISBN: 262-75105-4
ISSN: 0506-8347

CONTENTS

1. Banglanagar (formerly called the Second Capitol of Pakistan), Dacca, East Pakistan (Bangladesh), Louis I. Kahn. Substructure vaulting of the Presidential Square entry plaza under construction, 1965.

Editor's Introduction

Paula Behrens

THE BUILDING OF ARCHITECTURE is a step in the process that links metaphysical ideas with physical reality. For this issue of VIA, we have sought contributions that explore this process of building architecture. The expression of architectural intentions, first in drawings and then through the joining of materials, is a cyclical process in which theoretical concepts and pragmatic experience enrich each other. Thus, the physical making of architecture can be seen as an essential part of the search for meaningful form. The articles we have collected here are intended to demonstrate aspects of this search.

The root of the German word for architecture, *Baukunst*, is "building-art," which reveals the duality of an architect's work. While architecture is often discussed as if it were either the art of building or the building of art, both aspects of its nature can and should exist in tandem. As Louis Kahn described it, a great building, "must begin with the immeasurable, must go through measurable means when it is being designed, and in the end must be immeasurable."[1] He further described the "measurable act" as "the design, the making of things," which extends what is commonly regarded as the design phase all the way through the building phase. The immeasurable spirit, then, becomes enriched and in fact achieved by understanding and exploiting the measurable means. If architecture is the making of space, materials and technology have to be called into action to form the container of that space. The interplay between dreams and physical reality becomes a continuous process of synthesis and expression.

Of course, the contemporary architect is rarely involved in the physical act of building. His or her ideas and intentions are set down so that others may complete the construction. This separation of the mental and physical labors accentuates the separation of the building process from design, and makes it more difficult for the act of building to be integrated into design. The expression of personal vision is more easily seen in drawings since they are produced directly by the designer. The modern construction process, involving the contributions of bankers, developers, construction managers, labor unions, lawyers, etc., is not nearly as poetic or personal an enterprise. While the surge of interest in fine architectural drawings can be understood, in part, as a reaction to and a frustration with the "construction industry," the necessity for architecture to be built in order to be perceived spatially, temporally, and metaphysically is also the ultimate opportunity for architectural visions to be enlarged to include the challenge of the physical world.

As students, we are even more removed from the building process than many architects. Our training specifically focuses on conceptual images rather than on physical realities. This issue of VIA is an effort to balance that orientation by emphasizing the aspects of architecture that deal with its making. Nonetheless, we do not offer here a manual of building practices, but rather a collection of articles that demonstrate a broad concern for integrating concepts with construction. Participants with different perspectives—architects, client, contractor—have shared their perceptions and projects, while historians and theorists offer background and theoretical analysis on the topic.

The architects, Emilio Ambasz, Herman Hertzberger, Cesar Pelli, and David Polk, discuss projects ranging from corporate office towers to a house renovation in terms of the processes they used to realize their designs. Their presentations reveal that each architect proceeds in a different way and adjusts the process to fit the conditions of each new task. The process is not only complex, it is never repeated in exactly the same way.

The client, Abdullah Nuridin Durkee, the director of a small religious foundation in New Mexico, sought out the Egyptian architect Hassan Fathy and learned an ancient mud-brick building technology from his craftsmen, thus joining the small but active self-help movement which strives to continue the traditional unity of designer/builder/dweller.

The contractor, A. T. Seymour III of Thos. S. Byrne, Inc., reveals some little-discussed points of a much-discussed building, the Kimbell Art Museum, by quantifying its construction into yards of concrete and boatloads of marble, as well as describing the intricate technical procedures required for the construction.

The building practices of earlier times are explored by historians. The description of the construction system of Song Dynasty China by Else Glahn reveals a highly standardized, proportional system that responded to the varied requirements of climate, site and materials as well as to religious and social concerns. George Thomas presents the innovative use of concrete in the construction of a hotel in Atlantic City in 1914 which demonstrates the interplay between the potential of a material and the vision of the designer.

The modern age has seen radical changes in ways of thinking about architectural form as well as in the materials and technologies available. Alexander Tzonis and Liane Lefaivre take a broad look at the theoretical basis for modern theories of functionalism and "machine-like" aesthetics by examining the development of mechanization and its relation to architectural thinking in Europe since the 1700's. The concepts of truth and lie as moral determinants of form became significant issues as the modern age took hold and Karsten Harries explores this issue from the time of John Ruskin up through the work of Robert Venturi and Michael Graves. Marco Frascari's investigation uses examples from Alberti, Lodoli, and the work of Carlos Scarpa to discuss the significance of the architectural detail as a generator of architectural form.

Also presented is an edited version of a dialogue by the French Symbolist poet Paul Valéry. This work, which was apparently read by Le Corbusier as well as by Kahn, examines the psyche of the architect and explores the difference between the acts of thinking and making. It reveals an attitude toward the subject that while thoroughly connected to its historical context also offers insights to the modern reader. The commentary by Jill Stoner which follows the dialogue looks at the meanings that architects today can find in the work.

"What you should try to accomplish is built meaning. So get close to the meaning and build."[2] Aldo van Eyck's advice of twenty years ago is still relevant today. A concern for meaning in architecture must be carried through in an attention to its building. We hope that this volume explores some of the ways in which built meaning can be achieved.

NOTES

1. This thought has been recorded in a number of different versions. It can be found in the "Voices of America" Forum Lectures, a series on "Modern American Architecture" in 1960, originally entitled "Structure and Form."

2. A. Smithson, ed., Team 10 Primer (London, 1968 and Cambridge, Ma., 1974).

ACKNOWLEDGMENTS

"Building" a journal is no less difficult than building architecture, but the ideas and encouragement of many people helped to bring this seventh issue of VIA into reality. A number of students aided in the early work–Jim Dart, Ruth Finch, Bob Petito, Sally Pinigis, Rick Votta–and others helped bring it to completion–Andrew Blanda, Erik Johnson, and Chris Macneal. The faculty and staff of the Graduate School of Fine Arts, including Peter McCleary, Adele Santos, Lee Copeland, and Anne Wetzel, all lent their support. Anne Reinke and Christine Lamb of MIT Press provided valuable information and encouragement. Former VIA editors Darl Rastorfer and Grey Read helped with advice on the pitfalls of editorship. The copy editors, Deborah Gorman and Louis Gross, were patient advisors and able technicians. Work on individual articles was done by Michael Peters ("Hassan Fathy in New Mexico") and Julie Iovine of Cesar Pelli's office ("Architectural Form and the Tradition of Building"). Encouragement and useful criticism was also received from Steve Ziegler, Deborah Cholet, and Emily Johnson.

Finally, and most importantly, the effort, ideas and patience of the authors was what made this work possible.

2. Banglanagar—the substructure vaulting of the Presidential Square entry plaza after completion.

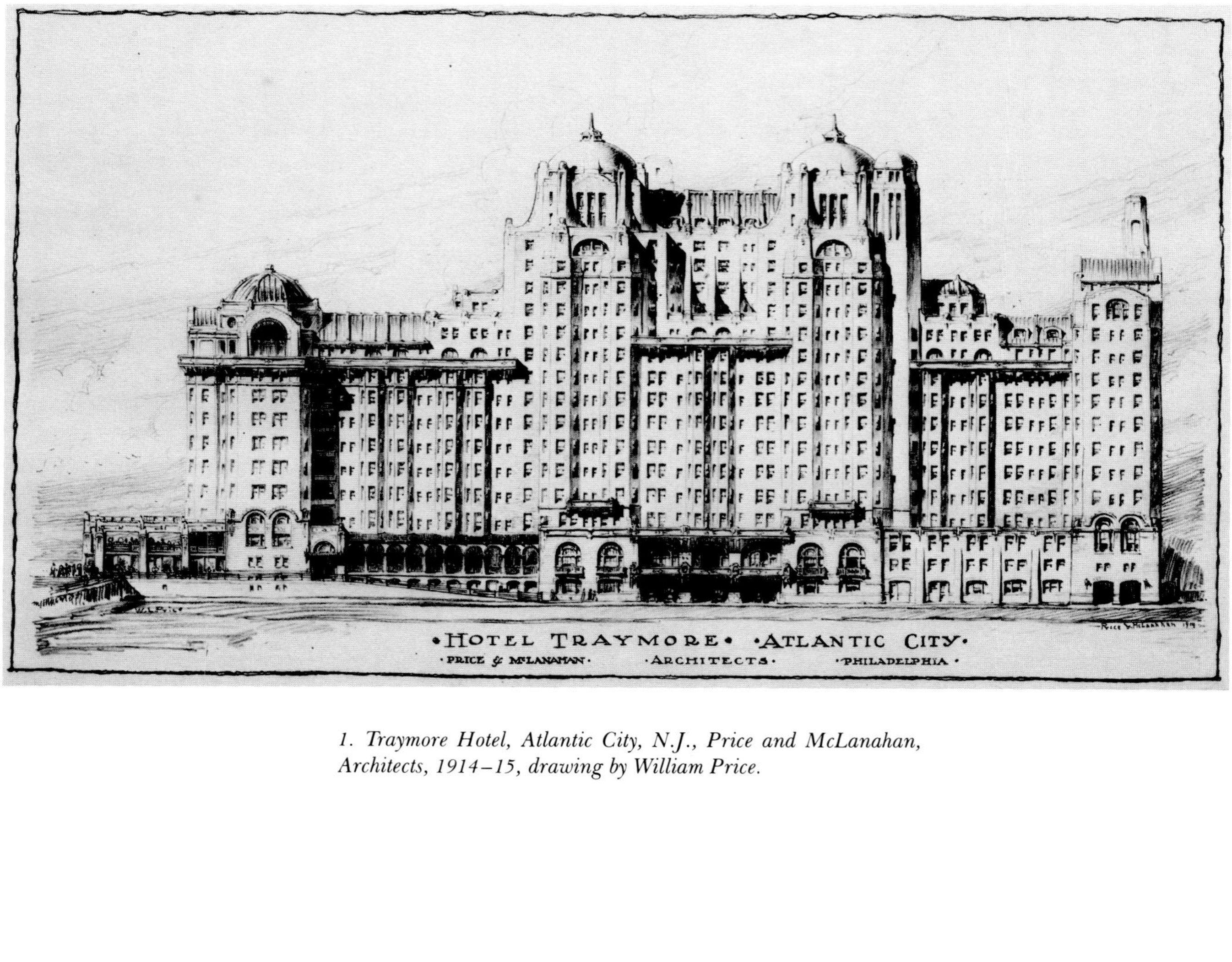

1. Traymore Hotel, Atlantic City, N.J., Price and McLanahan, Architects, 1914–15, drawing by William Price.

A HOUSE BUILT ON SAND

George Thomas

That the history of structures would preoccupy contemporary historians should come as no surprise, given the theoretical bent of the past century. Carl Condit's *American Building Art* and Peter Collins's *Concrete: Vision of a New Age* are both compelling achievements of historical synthesis that have done much to enable our generation to understand the sources, values, and standards of the principal inventors of advanced structures.[1] Unfortunately, with few exceptions, the most advanced structures have proven architecturally and aesthetically uninspired, while the more interesting architectural designs have tended to be structurally unimaginative. Ernst Ransom's work illustrates the former fact, while H. H. Richardson's, Frank Furness's and McKim, Mead, and White's fundamental failure to fully exploit structure has been an oft-told incident of architectural history.

Fortunately, in the early twentieth century there appeared a number of architects with lively architectural imaginations who sought to push the new structural systems of the industrial age to lengths heretofore unparalleled. Wright's early work in reinforced concrete has already been so extensively treated as to make it seem that imagination was solely the province of the Midwest. Certainly that was not the case in the aesthetic exploitation of reinforced concrete, for the East Coast produced the extraordinary group of reinforced concrete commercial buildings designed by Philadelphians Price and McLanahan. Beginning with the Jacob Reed's Sons store in Philadelphia (1903), the Blenheim and Traymore Hotels in Atlantic City (1905–06), and the Steubenville and Fort Wayne railroad stations of 1909–1910, Price's firm explored the technological and formal possibilities of the new material.[2] By 1914, when Price designed the five-hundred-room addition to the Traymore and the Chicago Freight Terminal (1914–16), he had demonstrated a clearly developed aesthetic notion about the material, which he had already articulated in a series of articles and lectures that spanned the last decade of his life.

It was this accomplishment that led George Howe to accord Price the status of the midwesterners Sullivan and Wright:

> America has always been the land of lone prophets without much but posthumous honor in their own country, which has been too busily engaged in exploiting its vast resources to take much interest in spiritual consequences of its own art, and Wright, Sullivan and Price were among the first to grasp the architectural possibilities of the new life and the new means of construction. Their names were known in Europe, while they remained comparatively obscure among their own countrymen.[3]

Despite Howe's enthusiasm, Price's work, far more than Sullivan's or Wright's, has been forgotten by architectural historians who have sought American roots for international modernism rather than accurately depicting the full range of American late Ruskinian design. Indeed, after Price's last principal publication, the only scholarly mention of his work since 1933 has been in reference to the structural achievements of the great concrete buildings in Atlantic City and Chicago.[4]

It is therefore appropriate, in a publication sponsored by the University of Pennsylvania, where Price himself lectured, to reevaluate Price's contribution to modern reinforced concrete. Such an opportunity exists because the largest and most complex of the concrete structures, Atlantic City's Traymore Hotel, was extensively documented in a photographic construction log that began in September of 1914 and continued until the opening of the hotel ten months later in the summer of 1915.[5] That construction log is augmented by a series of analytical articles in architectural and engineering journals that provide contemporary commentary on the building.[6] It therefore gives direct evidence on early twentieth century theory and practice while offering compelling testimony on the adventurous and creative designers who dared greatly—and overcame difficulties that overwhelmed better educated but more cautious followers.

Theory

Price's own views on the theory of concrete are strikingly illustrative of Ruskin's and Viollet-le-Duc's notions of the expression of material in construction, merged with a Hegelian belief that a building should be a reflection, like a cultural mirror, of contemporary society's values. Price's ideas can be readily assembled from his articles in turn-of-the-century publications. In 1906 he gave a lecture to the Association of Portland Cement Manufacturers that was published as their Bulletin No. 2 and was republished in the 31 March 1906 issue of the *American Architect and Building News* with illustrations of his recently completed Marlborough Blenheim Hotel.[7]

Price's essential understanding lay in his awareness that concrete construction tended toward unified structures in which column and beam or column and slab became essentially monolithic. His own view was that the entire building became a monolith when one made the frame a rigid truss, thereby avoiding problems of settling and cracking: "In concrete the building, if properly designed structurally, and properly built of proper materials, is a monolith."[8] Such a view is of course naive, but was in keeping with contemporary practices of both the Trussed Concrete Steel Company and the Hennebique Construction Company of New York, which had operated under similar assumptions as engineers for, respectively, the Blenheim Hotel in 1905 and the earlier Jacob Reed's Sons in Philadelphia of 1903. No expansion joints were provided, nor were the differential expansions of skin and structure accommodated. In the smaller Jacob Reed's building the problems were fewer; they were magnified as buildings grew larger.

2. Traymore Hotel after first addition, 1914.

Like Bogardus in regard to cast iron half a century earlier, Price presumed nearly supernatural strength for his chosen material: "It is my opinion that a considerable area of this building [the Blenheim] could have the foundation washed out without damage to the structure, nor does the heaviest wind so far encountered cause a perceptible vibration."[9] Such qualities, Price believed, deserved architectural expression. In so stating he was of course following the theoretical position of the mid-nineteenth century, expressed in England by John Ruskin and Owen Jones and in France by Viollet-le-Duc, that materials and methods of construction deserved specific representation. Since concrete was not of block construction but was rather a plastic and continuous material, it seemed to Price that different forms had to be developed.

> In block building, whether of stone, brick, or other material, the joints are an essential element of building, and as such should become an essential element of design, to be accentuated rather than hidden, and in all characteristic and good architectural design, you will find recognition of this fact. Take, for instance, any of the Classic orders, or their modification in the Renaissance, and you will find a frank succession of blocks . . . all built up, stone upon stone, with major joints marked by mouldings, bands, and ornament, on logically built structures. But if you attempt to follow such a system of design in a plastic material, or even a material like wood, the results are disastrous to true architecture. The design becomes a matter of external form, not of the true expression of methods and materials used. Concrete is built with a shovel and a trowel. . . .[10]

To Price, process and material gave concrete its inherent monolithic quality, which only continuous forms, and plastic shapes could denote. First in the Blenheim and then in the initial tower of the Traymore, the architect worked through shapes that avoided the right angles of block construction. One result was a characteristic detail of a curving junction where beam met column, which Price also carried over to the openings through masonry walls as a means of carrying the expression of the material to the exterior.

The question of exterior material and its decoration similarly preoccupied Price. In "The Decorative Treatment of Plaster Walls," Price argued for a "plaster," i.e., troweled Portland Cement surface over the concrete frame and the terra cotta tile infill panels. This produced a cool gray, monochromatic surface that visually represented the underlying material while emphasizing its monolithic quality.[11] But, the absence of control joints and expansion joints and the failure adequately to bond the concrete structure to the exterior caused the surface to fail—apparently within its first year, for Price's design for the neighboring Traymore Hotel fireproof tower, in 1906, provided for a yellow glazed terra cotta skin similar to modern clay piping, and with the same range of earth-toned colors.

The failure of the concrete skin over the concrete frame made it necessary for Price to switch to more traditional surface materials, many of the other themes of the Blenheim were continued in the later designs for concrete buildings. The location of the column was denoted in all of Price's later concrete buildings by pier caps breaking through the roof line above and by the projection from the pier of concrete brackets to support balconies, parapets, and other overhangs. Moreover, the form of the brackets was no longer based on historical precedent but, rather, was taken from the expression of the parabolic moment distribution, growing broader and deeper toward the columns. Form here became both a symbolic and a direct representation of function. Similarly, columns and beams directly reflected loads and spans without regard to historic detail or proportion.[12]

If the expression of structure reflected the neo-Gothic principles of the Frenchman Viollet-le-Duc, the use of ornament on concrete buildings was derived from the writings of the Englishmen: John Ruskin, Owen Jones, and William Morris. Architecture was, by its nature, ornamented construction, with ornament alone differentiating it from mere building. But, ornament was to be neither applied without principle nor freely invented. Ornament's purpose lay in accenting the constructive issues, as it had in Gothic design, and its forms were to be derived from the whimsical creations of a benevolent God.[13] Thus, naturalistic ornament derived from the marine setting decorated the Blenheim with terra cotta sea horses, giant crabs, lobsters, and sea shells framing openings, marking major changes of level, and proving that this was no ordinary urban hotel.

The choice of terra cotta was based on principle as well, for Price made it clear in his 1911 article that he viewed terra cotta as an analogue to concrete:

> If a closely allied material, which can be reasonably embedded in the wall surface be used in such a way as to seem a part of that surface there can be no objection to such use . . . ; and burnt clay products which can be fashioned in innumerable forms and colors, glazed and unglazed, when so separated in design as to allow the wall surface to penetrate and tie it to that surface is almost an ideal form of wall decoration.[14]

The choice of color, a lime green glaze, was also intended to represent the nature of the terra cotta as a molded surface but not as a structure: "The use of a moderate amount of terra cotta where required, for sills, copings and other wall protection has been made unusually important by modelling it in interesting and appropriate sea forms, and coloring it a light green, by a glaze which has the additional advantage of destroying any semblance of stone and giving it a plastic character that seems appropriate to modeled ornament."[15] Thus, like his principal Anglo-American contemporaries, Price sought sources for decoration in nature, and worked by analogy and context rather than from the abstract principles of International Modernism.

The Traymore Hotel

By 1914, when the major phase of the Traymore was underway, Price had modified but not significantly altered his design intention. Where the Blenheim decoration had been loquacious and narrative in its emphasis on seaside flora and fauna, the Traymore decoration was reduced to colored patterns of glazed tile set into pier surfaces, while Mercer tiles, ranging from plain to figurative pieces, were set into the concrete arch over the new boardwalk entrance. If there was less emphasis on naturalistic sources of ornament, the underlying theme of the vivid, gay, visually stunning Atlantic City boardwalk setting was fully developed in the gilded domes, multicolored pinnacled skyline and rippling surfaces of this modern Xanadu.

Still, themes enunciated by Ruskin and Viollet-le-Duc continued to be principal generators of architectural form. As in the Blenheim, wall surfaces were punctuated by a rhythmic line of pier caps that marked the structural frame, openings were arched to suggest the continuous nature of the reinforced concrete, curving transitions occurred between columns and beams, and the dominant architectural forms, the curving wall surfaces, barrel-vaulted roofs, and round domes all reflected the possibilities of poured plastic shapes. For Price, the purpose of ornament and architectural form lay in their capacity to represent function and to describe the construction process. Thus, like his twentieth century contemporaries in the mid and far west, Price provided a transition between the Gothic revival of the nineteenth century and progressive twentieth century structural and functional expression, but, unlike the designs of his European contemporaries, Price's commercial designs were successful in the democratic, mass culture world of modern America.

Price's explorations of reinforced concrete culminated in 1914 with the building of a five-hundred-room addition to the Traymore Hotel (Fig. 2).[16] Some

3. San Cristobal Hotel (project), Puerto Rico, Price and McLanahan, Architects, 1912, drawing by William Price.

4. Piles at the construction site, Traymore Hotel, 5 September 1914.

5. View of the site looking north, 8 October 1914.

of the difficulties encountered were by then commonplace though they may seem monumental to our age: the rapid completion of an immense building in the ten-month off season to capitalize on a two-month tourist season; the construction of a multileveled building with widely varying loads on sand; the design of a waterproof exterior to protect the building from the harsh seashore environment. Other problems were new. The architects were given the opportunity to be part owners if they provided plans and raised the necessary capital. With war clouding the financial markets, the architects closed financing the night before World War I broke out, and they ordered steel sections before U.S. industry shifted over to war production.[17] Of greater architectural interest were the problems of joining new concrete construction to the existing building. Price's scheme called for the complex interconnection of the old and new portions with multiple levels that necessitated the removal of existing floors. The 1906 tower was carried by temporarily bracing it while columns were replaced and new floor levels were grafted onto the old structure.[18] Finally, in a purely commercial environment, the hotel had to look like no other building on the Atlantic City beachfront.

Design drawings were completed in the summer of 1914, with the new building relating to the Pricean vocabulary that had evolved in a series of unexecuted projects between 1909 and 1914. Essentially, Price had decided that if concrete could not be the form, at least the plastic shapes inherent in a continuous and amorphous material could be represented. In the Overhanover, North Carolina, and the San Cristobal, Puerto Rico, hotel projects, the office had sketched rugged, clifflike masses, given order by the expression of the underlying concrete skeleton and by a "giant order" of bays that enlivened the façade (Fig. 3). The fenestrations on the lower levels and the great domes and barrel vaults on the upper levels hinted at the various functions of a modern hotel with vigor and imagination.

To Price and McLanahan's architectural sense of the materials were added the equally specific skills of the Philadelphia engineering firm of Otto Gentner and Dickinson Shaw.[19] In each of Price's earlier concrete projects, the design of the structure had been the responsibility of structural engineers outside the office, and this project was no exception.

The construction of the Traymore was undertaken by a Philadelphia company, Theodore Cramp and Company, which was better known for ship than for building construction. Cramp Ship had long been a fixture on the Delaware River, but in the early twentieth century it expanded to provide contracting services and work crews for reinforced concrete building. Though seemingly widely different in skills and crafts, the shipbuilding background was probably apparent in the organization of the site and in the use of long booms to move materials, especially concrete, from the central mixing machinery across the full expanse of the site.

Chronology

The old wood Traymore closed in late August and was demolished. By 5 September the site had been cleared, and the bark-covered pine pilings had been delivered to the site (Fig. 4).[20] The foundation excavations into the beach sand were shored with 2 × 10 wood planks, which were driven deeper as excavation proceeded. Eighteen-foot-long wood piles were driven with water jets into the sand and then tapped with a steam hammer to the point that they could no longer be moved (Fig. 5).[21] Piles were typically placed two feet, six inches on center, covered with a reinforced concrete mattress footing, and capped by a cast-iron column base to support the steel-covered columns typical in the mid-portion of the hotel. Foundations varied in size according to projected load.

While foundations proceeded on the north side, extensive demolition and reconstruction were occurring around the original tower. The one-story brick arcade was pulled off and the first two floors were removed and replaced to meet the levels of the new building. This work necessitated extensive temporary bracing while the new floors were being inserted; to receive the new floors, the 1906 columns were "shaved" down, and spiral reinforcing rods were cast into the concrete collars that spliced the new work to the old.[22] New balconies in the music room and vestibule were carried by steel brackets, stove-bolted to the columns, and tested with loads of 150 lbs. square foot (Fig. 6). During September and October the new three-story front ocean deck was constructed with reinforced concrete columns carrying slab and beam decks. Of particular interest was the then new "mushroom flat slab" floor of the exchange level garden, which framed the central fish pond that formed the ceiling of the "Submarine Grill" below. A photo of 8 October 1914 shows the orthogonal grid of 5/8″ reinforcing rods superimposed above the rectangular grid of similar bars (Fig. 7). To modern minds the absence of top steel suggests a simplistic notion of flat plate moments, but the results were apparently satisfactory.[23] By 13 November 1914 the front three-story addition (two above boardwalk level) was nearly complete, with reinforced concrete framing finished except for pouring the ribs of the main entrance's barrel vault (Fig. 8). Brick sheathing had reached the midpoint of the third floor, and falsework for the arched brick architectural openings was underway on the third floor. On New Year's Day, 1915, the front ornament was progressing, with sixteen men hard at work in the winter chill.

In the meantime, the foundation and excavation work of the extension continued through the early autumn. The organization of the site holds considerable interest for our era. A second immense concrete

6. Balcony test, old tower, first floor, 23 December 1914.

7. Steel reinforcing for mushroom flat slab, 8 October 1914.

8. Three-story boardwalk addition, 13 November 1914.

9. North mixer with concreting tower and hopper, 12 October 1914.

10. Rigged chutes, looking southeast toward first tower, 12 October 1914.

11. Erection of first steel columns for central tower, 24 October 1914.

mixer was erected on the north side of the site, beyond the midpoint of the new Illinois Avenue front, while two material towers were constructed at middle thirds of the long facade (Fig. 9). Pumps and a ten-story wood tower above the mixer transported mixed concrete to hoppers at various levels, from whence it flowed by gravity through a series of chutes carried by cables and rigged tripods (Fig. 10). Whether the rigging was the result of the Cramp Company's heritage as shipbuilders or had been transferred from older building technologies by the Edison Portland Cement Company remains to be determined. Whatever the source, the availability of the gravity feed distribution system guaranteed an unending flow of material that made the ultimate speed of the job possible. It should be noted that with the exceptions of the steam-powered mixers and hoists and of the steam-powered pile driver, which can be observed at various foundation positions, no other heavy machinery was in use on the job site.

The foundation, the boiler pit, and the waterproof basement slab were still under construction as winter began, but by 21 October the first steel columns of the central tower were being raised by the same tripods that guided the concrete chutes (Fig. 11). They framed the major corridors of the exchange floor and of the American- and European-plan dining rooms. The use of steel cores expedited construction by reducing curing time, and suggests that Price, like his contemporaries, had no theoretical commitment to absolute consistency of construction process. Indeed, the superstructure used both "hooped," or spirally reinforced, and "rodded" square section columns, as well as steel-cored columns. Beams were variously steel-cored or reinforced concrete. Typical floors were slab and beam, but long span tile was set between concrete joists over the mezzanine and first floors. The column layout was also consistent only in its many inconsistencies, with offsets from one floor to the next necessitated by the complex architectural

12. Pouring of second floor, 10 January 1915.

13. Formwork for fifth floor with spiral column reinforcing on deck, 26 March 1915.

form. Here, as in all of Price's buildings, the construction was the servant of the design, and not vice versa, as if in the process of discovering the limits of the new material, the architects set out to fully exploit every possibility and to use every available technology.

In the article on the construction of the hotel in the *Engineering News*, that complexity was noted: "Thus of the 179 columns starting at the foundation there are in addition 132 columns starting on girders at the various floors. Of the 41 columns that run up to the springing line of the dome only 9 continue to their own foundation. The 32 other columns are supported by girders at different floors."[24]

Though the construction systems were often exotic, once the basement had been finished, progress was meteoric. The exchange floor pours were completed by mid-December, the mezzanine forms were finished by New Year's Day, and a week later the first floor was ready for pouring and the American-plan dining room was under roof. Pouring began on the second floor on the eighth of January, 1915, and the fifth floor formwork was nearly complete two weeks later on the twenty-sixth (Figs. 12,13). During the harsh winter months, six stories were poured in twenty-four days. In another three weeks, the eighth floor was being poured, while the brick skin had already reached the top of the mezzanine (Fig. 14).

Although the speed seems breathtaking, it should be noted that the work was accomplished with reasonable precautions for the day. The wood forms remained in place long after each pour was completed, and straw was spread for curing and for protection from frost. The speed was attained largely through the size of the crew. At any one time forty to fifty men could be seen at work on various levels. When, a decade later, the builders of Schultz and Weaver's Breakers Hotel in Palm Springs crowed about the erection of this smaller hotel in nine months, they had nothing on the Traymore builders.

The exterior skin progressed at a similar pace.

14. Concreting eighth floor with brickwork to mezzanine, east side, 19 February 1915.

15. Concreting tenth floor with brickwork to fourth floor, east side, 5 March 1915.

16. *Ninth-story brackers and roof formwork, 9 March 1915.*

18. *North dome, 19 April 1915.*

On 19 February the exchange floor was sheathed and the falsework was in place for the first bedroom floor. Two weeks later, a view from the same position showed the brickwork up to the middle of the fourth floor and glazing beginning on the exchange level (Fig. 15). Complexities in the formwork of the upper floors slowed the pace, but by the ninth of March the ninth story balcony brackets were exposed, the ninth story was framed, and the roof formwork was ready to be poured (Fig. 16). By Easter Sunday, the essential shape of the hotel was clear, with only the skyline forms remaining to be constructed (Fig. 17). On April third and fourth the mid-Atlantic coast was hit by an immense snowstorm that dumped up to a foot of heavy, wet snow on the building. Despite the unexpected load, the Traymore stood its ground, a fact suggesting that safety margins were adequate.[25]

Topping Out

Obvious signs of progress nearly disappeared in mid-April while the forms were constructed for the setbacks and volutes that were to visually and physically counterbalance the intended domes (Fig. 18). Here it was obvious that the architectural forms were becoming more emblematic of the represented idea of structure than the thing itself. Columns did not necessarily line up with pier caps, nor did the volutes physically buttress the domes. Logic was intentionally sacrificed to create exuberant, overscaled forms that represented another level of public rooms, beginning at the twelfth floor with a ballroom, indicated by the great arched windows of the level, which in turn led to public rooms inside the domes (one of which served as a children's playroom, with murals from the brush of N. C. Wyeth illustrating children's classics, notably *Treasure Island*).

Those forms and setbacks were made possible by a daring 44-foot beam from which cantilevered brackets carried six columns of the thirteenth floor, while

17. *West side, 16 April 1915.*

19. Beam above twelfth floor Banquet Room with cantilevered brackets to carry thirteenth floor columns, 12 April 1915.

20. North dome, 20 April 1915.

portions of the eleventh floor were suspended from the same beam (Fig. 19). At the fifteenth floor, the domes were supported on diagonal beams resisted by the volutes carried on cantilevers from the twelfth floor. The domes were a hybrid form, with six ribs providing continuous integral support for the shell surface (Fig. 20). But, despite the ribs, the domes were intended to be self-sustaining, with concentric reinforcing and sufficient lap splicing to resist lateral thrust. However, the hybrid form has led historians of structures to speculate that the form was a ribbed shell, and not a true shell. In fact, when the hotel was demolished, the domes came off intact, a result demonstrating the accuracy of the original design.[26]

21. Building nearing completion, 19 June 1915.

Completion

By 19 June, less than three hundred days after the demolition of the old hotel, the building was essentially complete. The construction towers were being removed; scaffolding remained only for the final terra cotta decoration; the interior decoration was nearly finished, and curtains were in place in many of the guest rooms (Fig. 21). More than a half-million square feet of new construction had been completed, using new technology at a construction cost of less than two million dollars.

Even more notably, the project had been accomplished with relatively few problems. These few difficulties are only hinted at in the construction photographs. Certainly there were difficulties at foundation 93, which remained open long after the remainder of the basement was poured (Fig. 22). A separate account of the construction by the family of William Baldwin, superintendent of the site for Theodore Camp, recalled the workers' battling of "quicksand."[27] Perhaps such was the case at that spot. Eventually the column was moved to a different location. On 20 November column 83 was installed with a significant lean, but that lean was rapidly corrected. On 23 Janu-

22. Difficulties at foundation #93, 3 October 1914.

23. Repair of column #142, 29 January 1915.

24. Incomplete pour at column #72, 20 March 1915.

ary column 142 under the third floor slab was probably the victim of an incomplete pour, while icicles hanging from the slab above, shown in the photographs, give ample evidence of the cause of such a problem. The solution for splicing the columns of the original tower was used, with spiral reinforcing and new concrete collar (Fig. 23). Finally, on 20 March, an unusual Saturday photo session recorded column 72, poured without any concrete for more than five feet below the eleventh floor (Fig. 24). While the four corner 5/16″ rods are clearly visible in the photos, the remainder of the form is empty, though the beams of the eleventh floor are continuous. Otherwise, while there was evidence of honeycombing and segregation of the concrete from inadequate tamping, typical of the days before mechanical vibration of the forms, the work was satisfactory despite its speed. With its completion, reinforced concrete came of age as a modern building material capable of expressing the goals of modern accommodations for the public.

Epilogue

The hotel was an overwhelming success. Its Wyeth decorated children's playroom, its submarine grill below the fish tank ceiling, its great dining halls and lobbies provided the quintessential spaces for the democratization of leisure (Figs. 25, 26). But it was the rippling, multitowered façade, crowned by great gilded domes shimmering in the ocean air like a transported fragment of ancient Byzantium, that brought the crowds into the hotel's great spaces (Fig. 27). The architect's decision to design to the limit of the material rather than to limit the design to the simplest formwork was vindicated. Three years later, Price began the drawings for a forty-story addition to the north, intended to rise to a crowning dome surrounded by four lower domes.[28] The architect's death in the midst of the preliminary design ended the project, but not the hotel's popularity.

For the next half-century it was the Traymore that kept Price's reputation alive. The building was exhibited at the Paris Exhibition de L'Art Decoratif in 1925, and in Great Britain, where it was well received. In 1928 it was published in George Edgell's *The American Architecture of Today*, where it was described as the "picturesque seaside hotel, purged of its gimcrack and filigree, translated into the terms of stern modernism, with its picturesqueness preserved."[29] At the beginning of the next decade, a jury headed by Ralph Adams Cram, William Rutherford Mead, Raymond Hood, and Paul Cret selected it as the only resort building to be included in *Masterpieces of Architecture in the United States*. There it was analyzed by Louis Sullivan's chief advocate, Charles Harris Whitaker, as a survival of "the arts and graces that used to dwell with us in the less hurried life that once we knew," while its form grew "Sullivanlike" as "the natural sequence of an honest procedure."[30] Of greater importance, its

25. Music Room after completion.

26. Exchange Floor after completion.

27. View of completed building from west.

domed, tapered, chamfered, and sculpted forms shaped the aesthetic of the 1920s from the Nebraska State Capital to New York Art Deco. Indeed, few buildings in the commercial culture of America so thoroughly anticipated the taste of a generation as did the Traymore of 1914.

The importation of the principles of Franco-Prussian International Modernism doomed Art Deco, and with it, the Traymore. By the 1960s, it was mentioned only by Carl Condit for its structural daring but not for daring to push structure to the limits of high architectural drama. In 1972 the Traymore was destroyed, dynamited into physical oblivion by Controlled Demolition, Inc.

That demolition, however, provided final answers to many questions raised during the construction and by later historians. Failure to provide expansion joints or to adequately anchor the exterior masonry had caused significant failure of the skin. Large cracks in the skin had developed, and pointing had lost its effectiveness in keeping out water. Eventually, the owner painted the sand-colored brick and terra cotta with International Style white paint. Worse, the theoretical belief that the building was a monolith had caused the designer to make no provisions for expansion. As a result, the skeleton had cracked to create its own uncontrolled joints, which significantly weakened the fabric. Finally, the harsh salt environment and the probable use of salt water in the concrete to prevent freezing had wreaked havoc on the concrete-encased steel columns, many of which had rusted a foot or more up from the footing. At the point that the owners abandoned the hotel for tax reasons, the Traymore was nearing structural failure.[31]

But the problems of the Traymore as it neared its sixtieth year should not disguise the essential triumph of its design or its importance in the history of architecture and construction. It is significant that in 1914 the organization of the site, the technology of construction, and the use of the materials would not be unfamiliar to a contemporary construction crew. It is significant, on the other hand, that construction of reinforced concrete still remains largely a handcraft. This fact suggests that work processes, like artistic forms, have a life of their own. Unfortunately for our environment, the richness of architectural form attained in the Traymore has not had a similar survival.

NOTES

1. Carl W. Condit, *American Building Art*, 2 Vols., *The Twentieth Century* (New York: Oxford University Press, 1960–61). Peter Collins, *Concrete: Vision of a New Age* (New York: Horizon Press, 1959).

2. Those buildings are discussed at length in George E. Thomas,

"William L. Price (1861–1916), Builder of Men and of Buildings," Ph.D. diss., University of Pennsylvania, 1975; the Jacob Reed's Sons Store is described in George G. Morris, "Jacob Reed's Sons Store," in *Philadelphia: Three Centuries of American Art*, ed. Darrell Sewell (Philadelphia: Philadelphia Museum of Art, 1976), pp. 472–473. The Blenheim and Traymore Hotels were extensively written about. For the Blenheim, see H. Toler Booraem, "Architectural Expression in a New Material," *Architectural Record*, 23, no. 8 (Aug. 1908), pp. 249–268; The Traymore is reviewed in John J. Klaber, "America's Greatest Seaside Hotel–The Traymore, Atlantic City, New Jersey," *Architectural Forum*, 127 (November 1917), pp. 119–128.

3. George Howe, "What Is Modern Architecture Trying to Express," *American Architect*, 137 (May 1930), p. 24.

4. See Condit, *American Building Art*, Vol. 2, 158 and Collins, *Concrete*.

5. The 200 negatives of the construction log were taken by Haworth's Photographic Service of 1020 Chestnut Street, Philadelphia, using 120 size roll film. They were auctioned from materials in the possession of the McLanahan family and came into the author's possession in 1970.

6. The first block of the Traymore was reviewed in "The Construction of the Traymore Hotel," *Engineering News Record*, 54, no. 19 (10 November 1906), while the major article is "Largest Fireproof Resort Hotel Completed at Atlantic City," *Engineering Record*, 82, no. 1 (3 July 1915), pp. 18–23.

7. William L. Price, "The Possibilities of Concrete Construction from the Standpoint of Utility and Art," *American Architect and Building News*, 89, no. 1579 (31 March 1906), pp. 117–120; see also "The Decorative Treatment of Plaster Walls," *The Brickbuilder*, 20, no. 9 (Sept. 1911), pp. 181–184.

8. Price, "The Possibilities of Concrete Construction," p. 119.

9. Ibid., p. 119.

10. Ibid., p. 119. The appearance of this theme in Price's writing suggests an awareness of contemporary French criticism of Hennebique's design. See Collins, p. 70.

11. When Price shifted back to a block exterior, he left the more radical solution for the solution favored by the bulk of concrete designers. See Collins, p. 70.

12. That Price was aware of the revolutionary nature of his decision to directly reveal the concrete beams in the public spaces of the Blenheim and later of the Traymore is apparent from his comments in "Modern Architecture, Part II," *The Artsman*, 4, no. 3 (April 1907), pp. 274–5. "We had in the Exchange, concrete construction, which is absolutely expressed in the resultant forms in the ceilings, and where we had big beams there are still big beams and where the spans were smaller, they were smaller beams. The construction is honestly accepted, and it does not seem to be very shocking to the public, or to many of the architects who have seen it."

13. Price's view that what he was doing was a continuation, or better, an evolution rather than a revival of Gothic is apparent in his discussion of "Modern Architecture," *The Artsman* 4, no. 2 (January 1907), pp. 227–237.

14. Price, "The Decorative Treatment of Plaster Walls," pp. 181–182.

15. Ibid., p. 182.

16. The best assemblage of views of The Traymore occurs in Edward W. Hoak and Willis Church, *Masterpieces of Architecture in the United States* (New York: Charles Scribner's Sons, 1930), pp. 167–179. The Historical Society of Pennsylvania, Print Department, has several hundred blueprints and tracing paper sketches taken from the Hotel before its demolition in 1972.

17. The story of the financing comes from the autobiography of the lawyer, and Rose Valley resident L. Stauffer Oliver, *The Bench is a Hard Seat* (Philadelphia: Dorrance, 1967), pp. 6–8.

18. "Largest Fireproof Resort Hotel Completed at Atlantic City," *Engineering Record*, 72, no. 1 (3 July 1915), p. 18.

19. Though both men worked in Philadelphia, no professional accounts of Shaw's or Gentner's careers have been found. Record keeping for the engineering profession is considerably more haphazard than even that of their architectural peers.

20. The dates of the chronology come from the inked titles on the photographic construction log.

21. "Largest Fireproof Resort Hotel Completed," p. 20. Loads were calculated at 15 tons/pile to take into account the expected effect of the widely varying loads from three to seventeen stories.

22. The method of building of the original Traymore is described in "The Construction of The Traymore Hotel" (*Engineering News Record*). The reinforced concrete columns, followed the column and beam system, rather than the steel-cored columns of the later building. In an interview with Jack Loiseaux, of Controlled Demolition, Inc. of Towson, Maryland, he recalled the excellent condition of the older portion, while the later building's column cores were badly damaged by salt-induced rust.

23. The mushroom or column-and-slab form of construction had been patented six years earlier, in 1908, by Minneapolis engineer C. S. P. Turner, but they had not been adapted to public rooms prior to their use in the Submarine Grill of the Traymore. See Condit, 2, pp. 166–170.

24. "Largest Fireproof Hotel Completed," p. 21.

25. Ibid., p. 23.

The live loads used in the design were as follows:

	lb./sq. ft.
Roof	30
Sleeping floors	60
Laundry floor	150
Kitchen and exchange floor	100
Garden (plus dead load of earth fill)	200

Live loads were reduced as floor areas increased and most column loads were reduced as heights rose. Stresses were calculated according to the following chart:

	lb./sq. ft.
Concrete in direct compression	500
Concrete in compression underbendings	500
Concrete in shear	100
Concrete in bond to steel	75
Steel in tension	16,000
Steel in shear	10,000
Ratio modules of elasticity, concrete to steel	1 to R.

The allowable stresses seem conservative but, given construction quality, they were reasonable.

26. Condit, p. 158.

27. That problem came to light in a letter from the sister of William H. Baldwin, Anna McAllister to her grandnephew, dated 21 March 1972.

28. *New York Times*, 8 Sept. 1916, p. 18, col. 6.

29. George Edgell, *The American Architecture of Today* (New York: Charles Scribner's Sons, 1928), 340–342.

30. Charles Harris Whitaker, "The Hotel Traymore," in Hoak and Church, *Masterpieces of Architecture in the United States*, p. 167.

31. 1982 conversation with Jack Loiseaux of Control Demolition, Inc. The main columns had dropped a foot and more, cracking the slabs.

1. Ziggurat detail in Rosso Verona marble on façade of the Banca Popolare di Verona. Carlo Scarpa, 1974–1981.

THE TELL-THE-TALE DETAIL

Marco Frascari

THE ARCHITECTURAL COMMUNITY has traditionally ascribed the maxim "God lies in the detail" to Mies van der Rohe.[1] The German version of the adage, *Der liebe Gott steckt in Detail*, perhaps the original source of Mies's maxim, was used by Aby Warburg to indicate the foundation of the iconographical method for researching in art history. The French version has been attributed to Gustave Flaubert, and in this case the maxim indicates a manner of literary production.[2] The common denominator in these different forms and uses indicates that the detail expresses the process of signification; that is, the attaching of meanings to man-produced objects. The details are then the *locii* where knowledge is of an order in which the mind finds its own working, that is, *logos*.[3]

The aim of this paper is to indicate the role of details as generators, a role traditionally ascribed to the plan, and to show that technology, with its double-faced presence as "*technè of logos*" and "*logos of technè*,"[4] is the basis for the understanding of the role of details. That is to say the "construction" and the "construing" of architecture are both in the detail. Elusive in a traditional dimensional definition, the architectural detail can be defined as the union of construction, the result of the *logos* of *technè*, with construing, the result of the *technè* of *logos*.

Details are much more than subordinate elements; they can be regarded as the minimal units of signification in the architectural production of meanings. These units have been singled out in spatial cells or in elements of composition, in modules or in measures, in the alternating of void and solid, or in the relationship between inside and outside.[5] The suggestion that the detail is the minimal unit of production is more fruitful because of the double-faced role of technology, which unifies the tangible and the intangible of architecture. As Jean Labatut, a French Beaux-Arts–trained Princeton professor of architecture notes:

> Whatever the air spaces, areas and dimensions involved, it is the precise study and good execution of details which confirm architectural greatness. "The detail tells the tale."[6]

In the details are the possibilities of innovation and invention, and it is through these that architects can give harmony to the most uncommon and difficult or disorderly environment generated by a culture.[7] The notion that architecture is a result of the resolution, substitution and design of details has always been a latent concept in architects' minds. That is to say, there is truth in the classical commonplace of architectural criticism: "That might have been great architecture if only somebody had worked out the details. . . ." Careful detailing is the most important means for avoiding building failure, on both dimensions of the architectural profession–the ethical and the aesthetic. The art of detailing is really the joining of materials, elements, components and building parts in a functional and aesthetic manner. The complexity of this art of joining is such that a detail performing satisfactorily in one building may fail in another for very subtle reasons.[8]

The discussion of the role of detail in the architectural process of signification will be developed in two parts. These inquiries analyze the understanding of the role of the detail within two different but interlocking realms, the theoretical and the empirical.

The first part is a search for an understanding of the concept of details in different levels of architectural production. The result of this inquiry is the conceptual identification of the detail with the making of the joint and the recognition that details themselves can impose order on the whole through their own order. Consequently, the understanding and execution of details constitute the basic process by which the architectural practice and theories should be developed.

The second part is an analysis of the architecture of Carlo Scarpa (1906–1979), a Veneto architect. In Scarpa's architecture, as Louis Kahn pointed out, "detail is the adoration of nature." The architectural production of this architect, in which the adoration of the making of joints is almost obsessive, allows an empirical interpretation of the role of detail in the process of signification, seen within culturally definable modes of construction and construing. In Scarpa's works the relationships between the whole and the parts and the relationships between craftsmanship and draftsmanship allow a direct substantiating *in corpore vili* of the identity of the processes of perception and production, that is, the union of the construction with the construing in the making and use of details.

Dictionaries define "detail" as a small part in relation to a larger whole. In architecture this definition is contradictory, if not meaningless. A column is a detail as well as it is a larger whole, and a whole classical round temple is sometimes a detail, when it is a lantern on the top of a dome (Figs. 2,3). In architectural literature, columns and capitals are classified as details, but so are piani nobili, porches, and pergolas. The problem of scale and dimension in those classifications and the relationship between aediculas and edifices makes the dictionary definition useless in architecture. However, it is possible to observe that any architectural element defined as detail is always a joint. Details can be "material joints," as in the case of a capital, which is the connection between a column shaft and an architrave, or they can be "formal joints," as in the case of a porch, which is the connection between an interior and an exterior space. Details are then a direct result of the multifold reality of functions in architecture. They are the mediate or immediate expressions of the structure and the use of buildings.[9]

The etymological origin of the word "detail" does not help at all in understanding the architectural use of the term.[10] In architectural literature the term appeared in the French theoretical works of the eighteenth century and from France spread all over Europe. This spread was caused by the coupling of the term with the concept of "style" and by the active influence of French literary criticism and theory on the French neoclassical architects. In 1670 Despreaux Nicolas Boileau, in the first part of his *L'Art Poétique*, warning against the use of superfluous details in poems, set an analogy between an overdetailed palace and an overdetailed poem.[11] By the eighteenth century this analogy was a commonplace and, ascribing it to Montesquieu, Giovanni Battista Piranesi attacked it as trivial in his defense of his architectural theory of overdetailed buildings.[12]

The French theoreticians of the *architecture parlante* were the ones who formally consolidated the role of detail in architectural production. In the analogy of the "speaking architecture," the architectural details are seen as words composing a sentence. And, as the selection of words and style gives character to the sentence, in a similar way the selection of details and style gives character to a building. This powerful role of the detail as generator of the character of a building was also pointed out by John Soane in one of his lectures on architecture: "Too much attention can not be given to produce a distinct Character in every building, not only in great features, but in minor detail likewise; even a moulding, however diminutive, contributes to increase or lessen the Character of the assemblage of which it forms a part."[13]

In the Beaux Arts tradition the understanding of the role of detail as a generator of the character of buildings determined a very peculiar graphic means for the study of it, the *analytique*. In this graphic representation of a designed or surveyed building the details play the predominant role. They are composed in different scales in the attempt to single out the dialogue among the parts in the making of the text of the building. Sometimes the building as a whole is present in the drawing, and generally it is represented on a minuscule scale, and so it seems a detail among details. The origin of the *analytique* and its role in the construing of architecture can be traced back to the technique of graphic representation and composition developed by Piranesi in his etchings surveying the *Magnificenza* of Roman architecture (Fig. 4). These are a graphic interpretation, with a stronger Vichian bias, of Carlo Lodoli's understanding of the built environment as a sum of inadequate details to be substituted with more appropriate ones.[14] Another form of the *analytique*, illustrating the architecture of Italy, can be found on the back of Italian *lire* notes today (Fig. 5).

It is important to notice that the *analytique* as graphic analysis of details had its development in a period in which architects did not have to prepare working drawings showing the construction of the details. The drawings carried few if any details and dimensions. The designer could be almost entirely dependent on his craftsmen. Builders had no need for drawings to show details whose execution was a matter of common knowledge. Construction of details was parceled out among the various tradesmen, who supplied the necessary knowledge for making them. The same craftsmen who furnished the information for the Diderot and D'Alambert *Encyclopédie* were able to construct the drawing with the exact eye of the artist, and the *analytique* was simply the source for the understanding of the ordering role of a single

2. *Chicago Tribune Building competition entry by Paul Gerhardt, 1922.*

3. *Lantern and dome of Santa Maria del Fiore, Florence. Filippo Brunelleschi, 1420–34.*

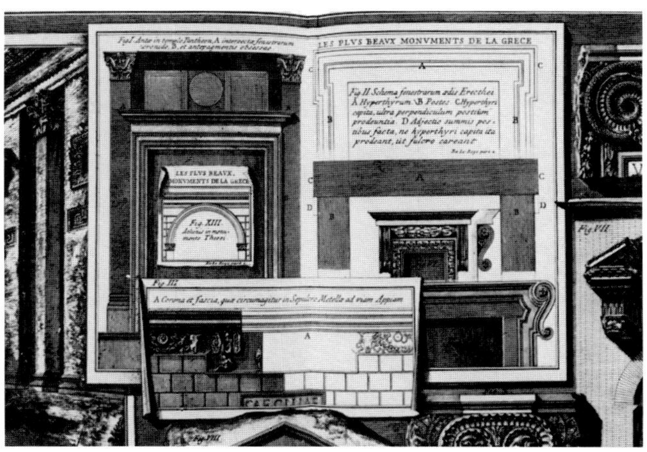

4. *Analytique from G. B. Piranesi's* Della Magnificenza dei Romani, *1761.*

5. *Italian 50,000 Lire note, a monetary analytique.*

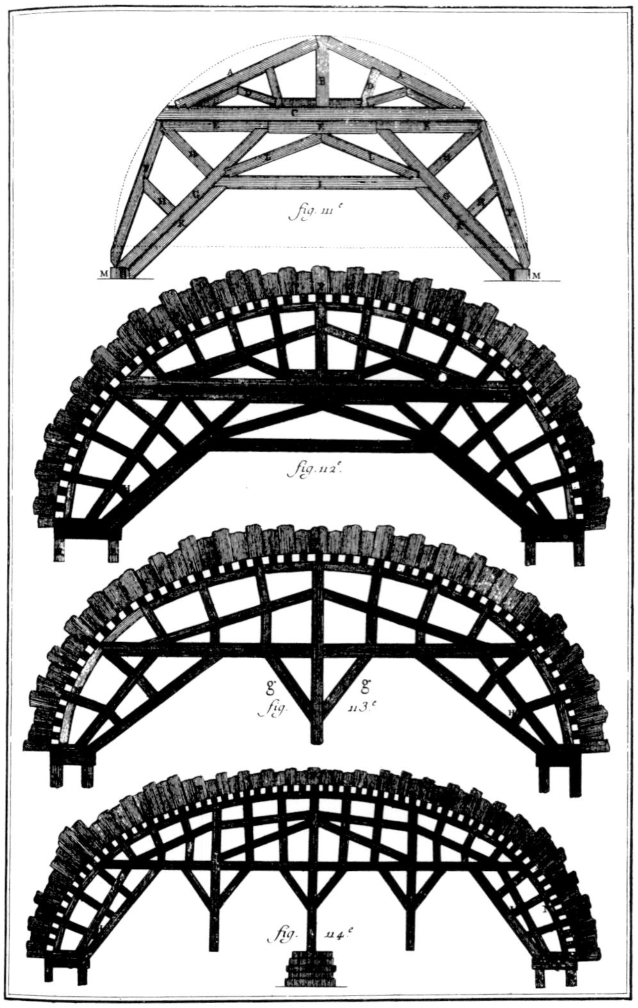

6. *Illustration of wood centering for stone vault construction from the* Encyclopédie *of Diderot and D'Alambert.*

detail in the overall composition (Fig. 6).[15]

The production of details, as it was established before the development of the industrial society and motivated by different cultural needs, began to become problematic in a predominantly economically motivated society. No longer considered as long-lasting cultural and social repositories, buildings came to be viewed as economic investments with an intentionally planned short existence. Two polar reactions had developed from the change that occurred in the scope of edifices. One of the reactions was that the various building trades no longer inferred the construction of the detail from design drawings. The details were studied and resolved on the drawing boards. Draftsmanship was substituted for workmanship, and the development of "real details" was replaced by "virtual" procedures. From this point of view the detail was no longer part of the building. The detail was no longer seen as a joint; instead, it was seen as a production drawing. In an *American Glossary of Building* the term "detail" is defined as "the delineation to full size or a large scale of any portion of an architectural design."[16] A French glossary was even more precise in this understanding of detail: "Detail: Specification or description of the work to be performed in the execution of a building."[17] In this interpretation "details" are verbal and graphic means for controlling the work of variable crews of vocationless workers who are unprepared for their own jobs and possibly even financially dishonest.

The second reaction to the change that occurred in the role of detail is the one that can be exemplified by the architecture produced by the Arts and Crafts Movement. The detail, in this movement, was seen as the means for the redemption of workers. The skill and knowledge of the making of details were given back to the workers. Workmanship was seen as the sole parameter for the details, which in themselves were seen as refinement of building tradition. The knowledge of details and of the related skills was the necessary means for the architect to practice his profession, since it was his task to select the appropriate workers for the appropriate details.

This duality in the physical production of detail is also found in the mental production. Using a conceptual analogy, it is possible to define architecture as a system in which there is a "total architecture," the plot, and a detailed architecture, the tale. The detailed architecture is based on "the constant process of drawing extrasystematic elements into the realm of the system and of expelling systematic elements into the area of not system. . . . The stone that the builders of a formed and stabilized system reject for being, from their point of view, superfluous and unnecessary, turns out to be the cornerstone of subsequent system."[18] From this point of view architecture becomes the art of appropriate selection of details in the devising of the tale. A plot with the appropriate details becomes a fully developed and successful "tale."

Architecture as art of the appropriate is the theme of Leon Battista Alberti's architectural theory. Alberti sees architecture as the art of the selection of appropriate details whose result is beauty, which is a meaningful goal. He defines beauty as "the 'concinnity' of all the details in the unity to which they belong"; in other words, beauty is the skillful joining of parts by a normative by which nothing can be added, subtracted, or altered for the worse. Generally this principle has been interpreted as stating that a building should be a complete and finished whole, a total architecture. Alberti, however, does not apply this concept to the actual edifice, but, rather, to the mental one.[19] The joint, that is, the detail, is the place of the meeting of the mental construing and of the actual construction. A perfect instance of this union of mental function and physical representation is in the façade of Palazzo Rucellai, designed by Alberti in Florence. Although the façade is incomplete and its incompleteness is clearly shown, the detailed architecture is complete, and nothing can be added or subtracted for the worse. The grooves of the joints of the stone slabs composing the thick veneer of the florentine *schiacciato* (representing the post and beam structure of the three superimposed classical Orders, related with arched windows and infilling walls) are the solution of the mathematical problem set by the relationships existing among the parts of the façade. In many cases the joints are not real ones, and the shapes of the stones are not as regular as they appear; fake grooves were carved in the stone to make the detailed architecture complete and to offer at the same time its own proof (Figs. 7,8).

Alberti's search for "Beauty" is the setting of a precise relationship between the detail and the attached meaning. Beauty is the result of the process of signification, and *concinnity* is the process for achieving it. Concinnity is the correspondence of three basic requirements: 1) Number, 2) Finishing, 3) Collocation.[20]

Number is a system of calculation. "The technique of calculation is part of the technique of house building."[21] Numbers in this way are tools for giving meaning. In architecture there are elements, and, in order to build, it is necessary to draw numerical correlation among them. In a triforium, three arches are correlated to four columns to make a serliana. The proof is in the details, and it is expressed in terms of mouldings, capitals, bases, and keystone. "Numerology," then, is for Alberti a technique for the selection of figures, thereby signalling that the details are related to memorable shapes such as the human body or cosmological figures.[22]

Finishing is a mathematical procedure for the definition of the dimensions of the directions in which the space of architectural objects is articulated. The edges of the tridimensional bodies of architecture are defined by a system of proportions. Proportion or "analogy" is the use of relations in a measurement.[23] An analogical system is a set of norms for the creation and combination of details. A basic measure, or module, is the norm from which all the lengths, widths, and heights are derived, and any single detail is measured after it. Then all the parts of the building will stand to each other in a direct and intelligible relationship. This relationship stands even when its form does not yet have a verbalized expression.

Collocation is the composing by place, that is, the functional placement of the details. The function in this case not only is limited to the practical and structural dimensions but it embodies, as well, historical and aesthetic dimensions.[24] The placing of details, then, is deeply related to the other two requirements: numbers and analogies. The detail in this manner is not defined by scale, but, rather, the scale is the tool for controlling it.

The geometrical and mathematical construction of the architectural detail is in no sense a technical question. The matter should be regarded as falling within the philosophical problem of the foundation of architecture or geometry, and ultimately within the

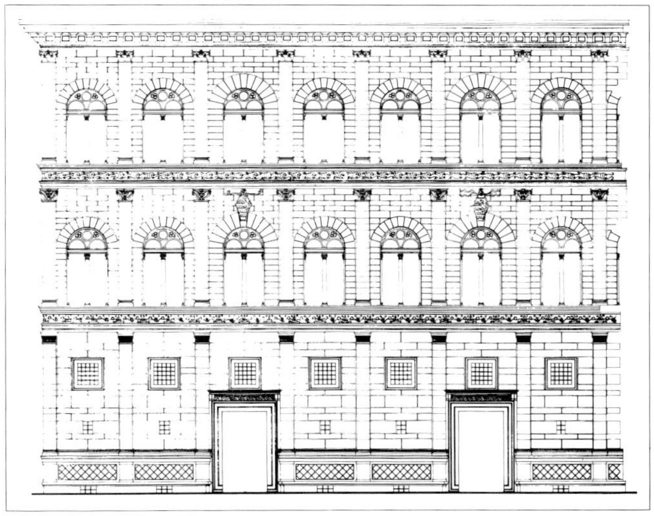

7. Palazzo Rucellai, Florence, showing apparent stone joints. Leon Battista Alberti, 1446–51.

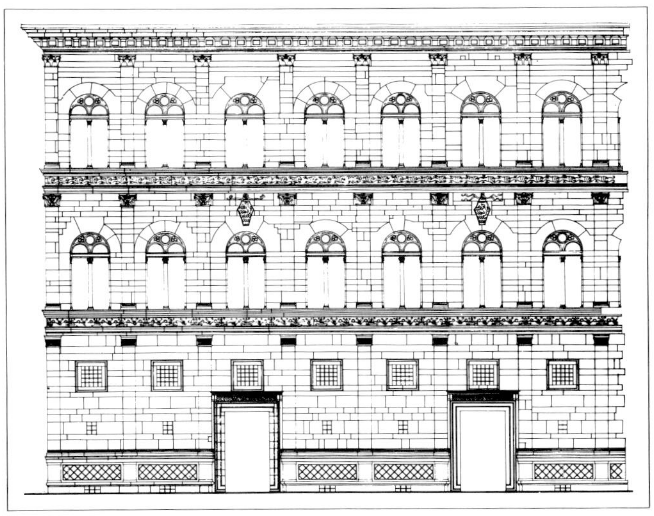

8. Palazzo Rucellai, showing the actual stone joints.

theories of perception.

The processes of designing, ordering materials, and building a house are techniques in the same way geometry is a technique by which the designer, the builder, and the user of a house transform the appropriate sign with a view to predicting the occurrence of certain events. This technique (geometry) provides us with a structure for describing the built world, a conceptual framework into which the designer, the builder, and the user can fit their empirical experience. Geometry shows how to derive a shape from another shape by transformation.

In this guise geometry does not state facts, but gives us the forms in which to state facts. It provides us with a linguistic or conceptual structure for the construction and the construing of a building. The geometrical structures embodied in the architectural details do not state facts but rather provide a structure for stating facts within a "scale." They give us a way of making comparisons that meaningfully relate

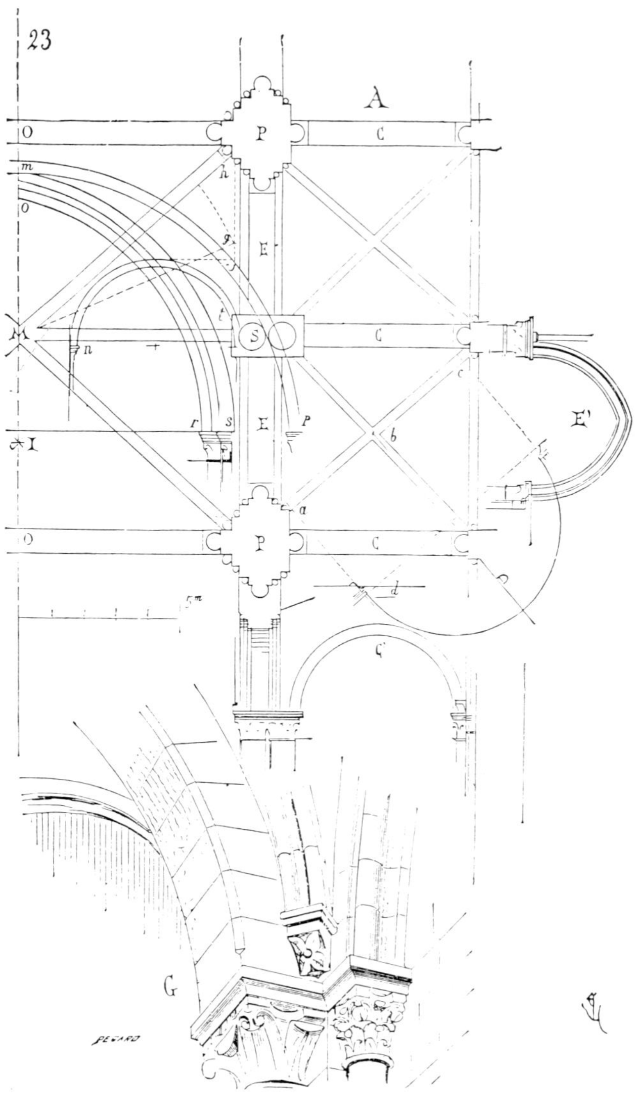

9. Study of the nave of Sens cathedral in the Dictionnaire Raisonné de L'Architecture *by E. Viollet Le Duc.*

visually perceived architectural details. The notion of the individually perceived details can be illustrated with the phenomenon of "indirect vision" as explained by Hermann von Helmholtz:

> The eye represents an optical instrument of a very large field of vision, but only a small very narrowly confined part of that field of vision produces clear images. The whole field corresponds to a drawing in which the most important part of the whole is carefully rendered but the surrounding is merely sketched, and sketched the more roughly the further it is removed from the main object. Thanks to the mobility of the eye, however, it is possible to examine carefully every point of the visual field in succession.[25]

Helmholtz's research on visual perception persuaded him that sensory stimuli only supply signs of the presence of architecture, but do not give us an adequate understanding of it. Such signs, that is, the details, acquire a meaning by virtue of which they become a vehicle of knowledge through a long process of association and comparison and through a set of geometric relationships.[26]

The geometrical relationships embodied in the details in a built environment as well as in a natural environment set the understanding of the large field of vision. The geometrical relationship or proposition at the base of the compound pier of the High Gothic architecture expresses in itself every feature of the imposed superstructure (Fig. 9). Such relationships are the results of the transformation in stone of the second requirement of Scholastic writing, of an "arrangement according to a system of homologous parts and parts of parts."[27] The details in this way, while forming an indivisible whole, are individually perceived and understood.

The problem of perception of details within the sphere of architectural appropriation is stated by Walter Benjamin: "Buildings are appropriated in a twofold manner: by use and by perception or rather by touch and sight. . . . Tactile appropriation is accomplished not so much by attention as by habit. As regards architecture, habit determines to a large extent even optical reception."[28]

This is an empirical theory that regards all perception of space as depending upon conventions and takes not only qualities, but even details as nothing more than signs, the meanings of which are learned only by experience. These conventions are the basis for architecture understood as existence, form, and location of external objects. These Helmholtz calls perceptions.[29] Perceptions are the ideas or signs of objects resulting from an interpretation of sensations that is carried out by processes of unconscious geometrical inference. The placing of details has a key role in these processes of inference. The visual sensations guided by the tactile sensations are the generator of the geometrical propositions. In architecture, feeling a handrail, walking up steps or between walls, turning a corner, and noting the sitting of a beam in a wall, are coordinated elements of visual and tactile sensations. The location of those details gives birth to the conventions that tie a meaning to a perception. The conception of the architectural space achieved in this way is the result of the association of the visual images of details, gained through the phenomenon of indirect vision, with the geometrical proposition embodied in forms, dimensions, and location, developed by touching and by walking through buildings.

The art of detail is in its most sophisticated and learned form in the work of Carlo Scarpa. An analysis of the concept of detail in Scarpa's architecture can best be begun with the words of Louis Kahn:

> In the work of Carlo Scarpa
> Beauty
> the first sense

Art
the first word
then wonder
then the inner realization of Form
the sense of wholeness of inseparable elements.
Design consults Nature
to give presence to the elements.
A work of art makes manifest
the wholeness of Form of the
symphony of the selected shapes
of the elements.
In the elements
the joint inspires ornament, its
celebration.
The detail is the adoration of
Nature.[30]

The "adoration of the joint," in Scarpa's architecture, is a perfect realization of Alberti's concinnity. Each detail tells us the story of its making, of its placing, and of its dimensioning. The selection of the appropriate details is the result of singling out its functional roles. The details of Scarpa's architecture solve not only practical functions, but also historical, social, and individual functions.[31]

Scarpa's architecture can be generically classified as the merging of the principles of the organic architecture as expressed by Frank Lloyd Wright with a learned distilling of Veneto craftsmanship with a blend modern and ancient technologies (Fig. 10). However, the definition is inadequate; whereas Scarpa's understanding of Wright's architecture was passive, based on an appreciation of photographs and drawings, his understanding of Veneto craftsmanship was active, based on his daily working and dealing with the stonecutters, masons, carpenters, glassmakers, and smiths of Venice. The result is a modern architecture that is more than rational structures and functional spaces. The teaching of functionalism is present in Scarpa's work, but the functionality is mediated by the search for representation and expression through the making. Scarpa's architecture stands against the bare structure of logic; it stands for the union of *res* and *verba*, that is, for the union of representation and function. This concept rules Scarpa's architecture from structure to expression. In his architectural objects the *technè* of the *logos*, the construing, becomes the manner of production of signs that are the details. The *logos* of the *technè*, the constructing, which results from the expression of Veneto craftsmanship, becomes the dialectical counterpart in the generation of the details as signs. Scarpa's buildings show indeed a constant search set between the actual form (the built one), and the virtual form (the perceived one). The constant manipulation of the discrepancies between virtual and actual forms is the method used for achieving expression. "In architecture," Scarpa once said, "there is no such thing as a good idea. There is only expression."[32]

10. Project for the Masieri Foundation, Venice. Frank Lloyd Wright, 1953.

The analysis of Scarpa's detail can be satisfactorily managed visually only by a continuous comparison between drawings and built objects, on the one hand, and the historical, practical, and formal reference that generated any single detail, on the other. It is also necessary to see Scarpa's details from two different sides. On one side, his detailing is the result of interfacing of design and craftsmanship on the site and of the constant "sensorial verification" of details during the assembly of the building. Scarpa made a practice of visiting the building site during the night for verification with a flashlight, thereby controlling the execution and the expression of the details. In the normal daylight it would indeed be impossible to focus on details in such a selective manner. It is also a procedure by which the phenomenon of the indirect vision becomes an element in the process of decision in the design. The flashlight is a tool by which is achieved an analog of both the process of vision and the eye's movement in its perception field (with only one spot in focus and the eye darting around). Another Veneto architect, Piranesi, used the same technique in visiting the sites of the buildings he was going to survey and represent in his etchings of the *Antichita Romane*. To single out the "expression of the fragments," that is, the details, he used the light of a candle.[33]

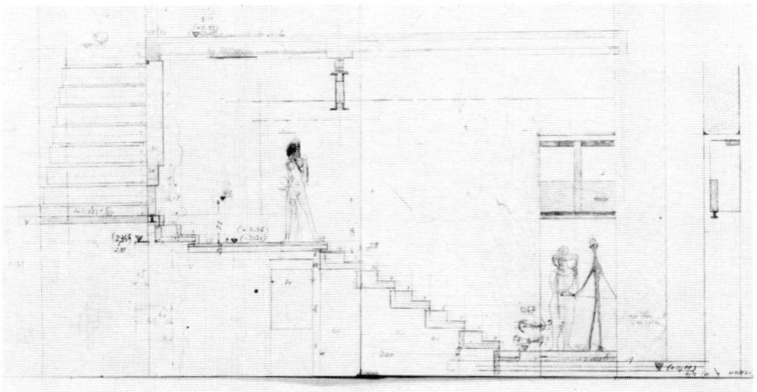

11. Stair at entry, Museo di Castelvecchio, Verona. Carlo Scarpa, 1958.

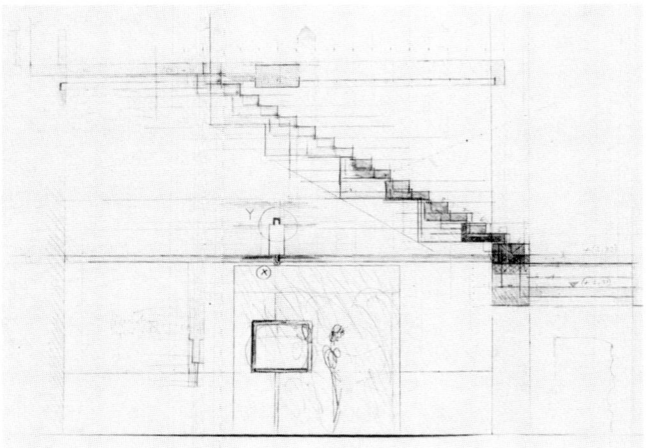

12. Stair detail, Castelvecchio. Carlo Scarpa, 1958.

13. Palace, Mantua. Fifteenth century.

On the other side, Scarpa's details are the result of an intellectual game performed on the "working drawings" that are the result of the interfacing of design and draftsmanship. That game is the matching of the construction of a representation with a construction of an edifice. The relationship between architectural drawings and buildings is generally thought of as a Cartesian representation based on visual matching of lines. However, Scarpa's drawings show the real nature of architectural drawings, that is, the fact that they are representations that are the results of constructions (Figs. 11,12). They are a construing of perceptual judgments interfaced with the real process of physical construction of an architectural object. The lines, the marks on the paper, are a transformation from one system of representation to another. They are a transformation of appropriate signs with a view to the predicting of certain architectural events, that is, on the one hand the phenomena of construction and the transformation by the builders, and on the other hand, the phenomena of construing and the transformation by the possible users. Consequently, on the same drawing there are present several layers of thought.

A design is developed by the same technique in which the drawing is made. The continuous inference process on which the design process is based is transformed in a sequence of marks on paper that are an analog for the processes of construction and construing. The piece of drawing paper selected for supporting the slow process of the construction of a design presents concurrently vertical and horizontal sections, as well as elevations of the designed piece. These drawings are surrounded by unframed vignettes that analyze tridimensionally any joint of the object, as in a prediction of the role of each detail in generating the whole text and in the perception of them in the "indirect vision." Scarpa's drawings do not define future architectural pieces as a simple sum of lines, surfaces, and volumes. Rather they present the process of transformation of the details from one system of representation to another, from drawing to building.

In Scarpa's drawings it is also possible to have the "proof" of the system of appropriation that rules the perception of architecture. These representations of three-dimensional structures on a two-dimensional surface result from the interaction existing between visual and tactile perceptions. The central part of the drawings generally presents graphic constructions that might be labeled a technical drawing. But they are not what are traditionally identified as plans, sections, and elevations. Scarpa's drawings are not merely devices of Cartesian descriptive geometry; rather, they are descriptions of the future perception in relationship to the making of the architectural object. The visual components of perception are analyzed for a detail and not for the whole, whereas the tactile perceptions are verified for the whole. These drawings present components that are not visible but that are the result and the projection of construction and construing, Alberti's mental edifice. They are the result of the memory effects of the organs of touch and sight in the making and using of architecture. These drawings are never fully rendered. Only fragments and parts of them are. This practice shows by analogy that, while it is whole, Scarpa's architecture cannot be characterized as complete. An architectural whole is seen as a phenomenon composed by details unified by a "device," a structuring principle. This principle, in Scarpa's architecture, is the order generated by the use and the understanding of classical architectural ideas such as façade design (Fig. 13).[34]

Scarpa is a *Magister Ludi*, and his buildings are texts wherein the details are the minimal unit of signification. The joints between different materials and shapes and spaces are pretexts for generating texts. The interfacing of commentaries with preceding texts in the architecture of Scarpa is always a problem of joints, and in the joint he achieves the change of conventions. That possibility is a consequence of the fact that many of his architectural texts are learned commentary to preceding texts and in many cases, as in a medieval *scholium*, the commentary in its interfacing with the original text is generating a new text. In the design of the addition to Gipsoteca Canoviana in Possagno, Scarpa was able to change the convention that asks for the background walls of a collection of gypsum casts to be tinted. Scarpa's solution was to put the white casts against a white background wall that was washed with light, without directly lighting the casts. The problem and the solution are in the use of light (Fig. 14). Scarpa solves it in a detail in the joint of three walls in a corner made of glass (Fig. 15,16). In a lecture given at the University of Venice (1976) he described the architectural making of this corner. The achievement of the effect of light occurs by a formal manipulation. The solution of the formal cause solves the final cause. He described it as "clipping off the blue of the sky," a formal cause, but the result was the lighting of the wall, the final cause. His own words are the best description of the making of an architectural detail:

> I love a lot of . . . natural light: I wanted to clip off the blue of the sky. Then what I wanted was an upper glass recess. . . . The glass corner becomes a blue block pushed up and inside [the building], the light illuminates all the four walls. My bias for formal solutions made me prefer an absolute transparency. Consequently I did not want the corner of glass to tie in a frame. It had been a *tour de force* because it was not possible to obtain this idea of pure transparency. When I overlap the glasses I see the corner anyway, especially if the glass is thick. One may as well put in the frame. Then, besides this, if it is a clear day one may see the reflection. Look, when I saw

14. Gallery, Gipsoteca Canoviana. Carlo Scarpa, 1957.

15. Window, interior view. Gipsoteca Canoviana.

16. Window, exterior view. Gipsoteca Canoviana.

> the reflection I hated myself. I did not think of it. These are mistakes which one makes in thinking, acting, and making, and therefore [it] is necessary to have a double mind, a triple mind, the mind like that of a robber, a man who speculates, who would like to rob a bank, and it is necessary to have that which I call wit, an attentive tension toward understanding all that is happening.[35]

17. Ziggurat detail at Brion-Vega Cemetery, S. Vito d'Altivole. Carlo Scarpa, 1970–

The development of architecture in the works designed by Scarpa proceeds by steps and stages. These are in the details. Each detail represents an interim result that cannot be considered a final result. Scarpa would invent details the precise architectural functions of which would become clear only after they had been used in several different designs. The range of those architectural functions goes from the immediate to the mediate understanding of the meaning of the detail. This creative use of details in design is fully in accordance with Wittgenstein's understanding of a creative use of language. The "exact" meaning, that is, the function of words, would only become known by a later use. A function of detail in a design becomes clear by re-presentation, that is, by re-use. The detail often appears incomplete and vague in its structuring principle. But, unifying in itself function and representation, the re-use of a detail becomes a creative catalyst. It becomes a fertile detail. The re-

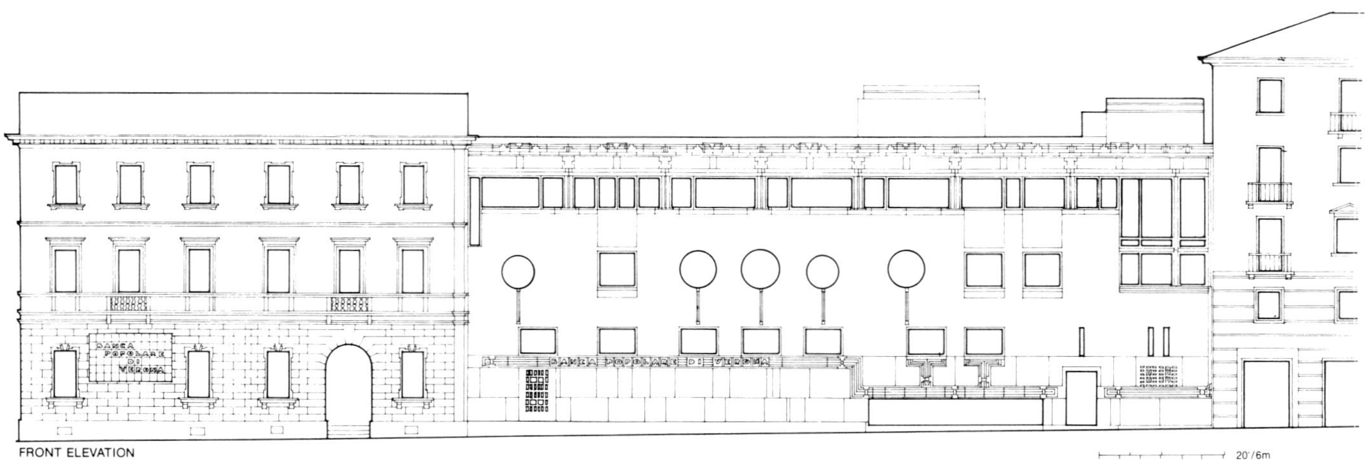

18. Façade, Banca Popolare di Verona. Carlo Scarpa, 1974–81.

use of details is analogous to Richard Wagner's re-use of *leitmotivs*.[36] The leitmotivs are structural devices used by Wagner to assemble and reconstruct the architecture of opera from within and are the smallest units of signification in the musical text. Scarpa's details are structural devices used to assemble architectural text from within.

A case of fertile details in Scarpa's architecture is the use of the "ziggurat" motif. The architectural function of these *fertile* details emerge in the Brion Cemetery at S. Vito d'Altivole (Fig. 17) and in the façade and the interiors of the Banca Popolare di Verona (Fig. 18). In the Cemetery, the ziggurat is executed in cast concrete, and it is a celebration of the possibility of casting as generator of moldings. In the bank, especially in the façade, the ziggurat detail is a *prima donna* in Rosso Verona, the brocade-like local red marble in which it is executed (Figs. 1,19–21).

Scarpa's first use of this detail was a cosmetic treatment of a temporary façade executed by piling up concrete blocks in front of the Italian Pavilion at the 1962 Biennale in Venice. But as Heraclitus has pointed out, the primary root of "cosmesis" is "cosmos." This same cosmetic detail becomes the principle of order in Scarpa's Museum of Castelvecchio in Verona. The ziggurat motif becomes the solution for terminating the layers of the wall of the façade to show the virtual joint between the original walls and

20

19

21

19–21. Ziggurat details, Banca Popolare di Verona.

22. *Joint, with Cangrande statue. Museo di Castelvecchio, Verona.*

24

23

25

23–25. *Cangrande statue, Museo di Castelvecchio.*

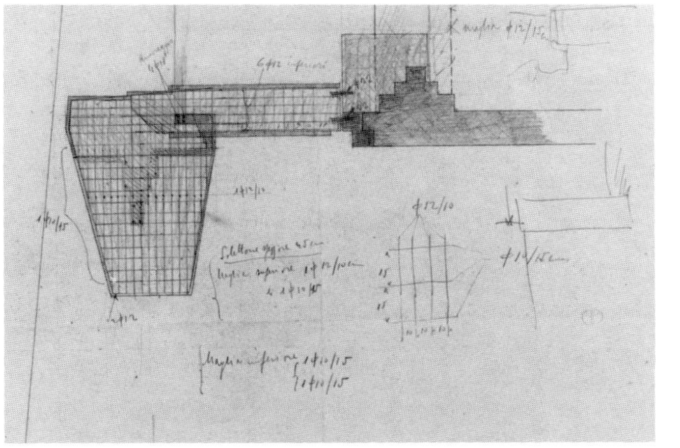

26. *Early design, platform for Cangrande statue, Castelvecchio.*

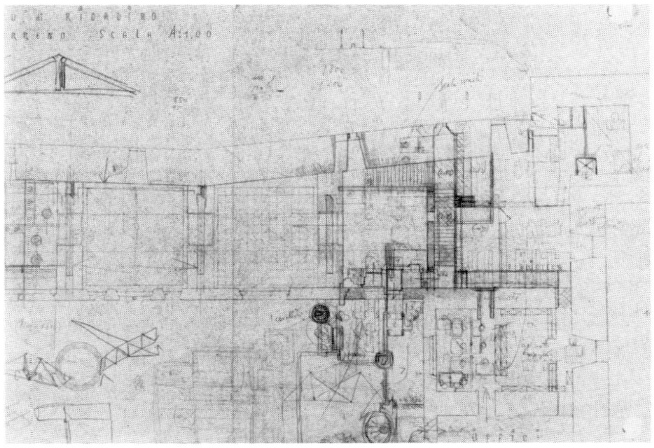

27. *Plan of entry, Castelvecchio.*

the Romantic replica of the façade wall constructed by Antonio Avena in 1924. In the Museum of Castelvecchio, the medieval equestrian statue of Cangrande and the structure which supports it are set in a spatial location that allows a view from the balcony, the bridge, and the court below. This location allows one to view the statue from close-up as well as from below, as it was seen in its original location on Cangrande shrine. This joint originates the full text of the spatial organization of the Castelvecchio Museum. It thus becomes the cause for the formal solution of the museum and the text in the con-text (Figs. 22–25).

An early design of the platform holding the statue of Cangrande shows it as the pretext for a celebration of the virtual joint determined by its collocation (Fig. 26). This drawing shows the idea of the ziggurat as a generator of the wall. The layers of the wall become independent units and each one of them is expressed in a vertical ziggurat. The space opened up by the cutting of the façade wall helps the whole composition of the new arrangement of the museum devised by Scarpa. The space, a virtual joint, is then the key articulation in the museum's path, but at the same time becomes a 'negative joint' in the articulation of the masses of Castelvecchio. The open space, instead of separating, helps connect the left and the right masses of the castle. These are situated on the sides of the tower which articulates the joint between the bridge on the river Adige and the castle. The selection of the ziggurat as the ending of the wall mediates the transition between inside and outside of the articulation. It exposes the materials of this complex architectural hinge composed of vertical planes defined by their framing relationship with the statue of Cangrande, the visual pin of the hinge. The ziggurat detail is also used in many other parts of the museum. In a study plan of the entry, this fertile detail is used in solving the joining of the stones used for the floor as well as in solving the deep reveal of the windows in the thick medieval wall (Fig. 27).

The ziggurat detail is also used extensively in the Brion Cemetery. The material, cast in place concrete, gives new meaning to the detail. The interaction between form and material moves the fertile detail from the realm of a production *sub specie utilitatis* to a production of *sub specie aeternitatis*. It is construed as a "ruin" loaded with memories before time. It becomes a perfect detail for the architecture of a cemetery, a place of memories. In this use the ziggurat finds its proof of being a fertile detail. A detail proves its fertility when it moves out of a private architectural language and becomes available through a collective production. A famous case of this is the Serliana Window that after being used by Palladio became a standard detail known as the Palladian Window. The Scarpian Ziggurat has indeed been used by many architects in their designs, but now is used in collective architecture. It has become a standard detail of Veneto cemetery architecture. The neoclassical temple in antis which has been the type for many family chapels has

28–29. Family chapels using ziggurat motif, San Vito d'Altivole, 1983.

been modified by a new model reference. The detailing of the Tuscan or Doric Orders has been replaced by a new detailing, a concrete cast-in-place Scarpian Ziggurat, a New Order (Figs. 28–29).

To conclude this discussion on the role of detail as a minimal unit in the process of signification (that is, the manipulation of meaning), it is useful to restate that architecture is an art as well as a profession. This is because of the understanding generated by the detail as joint. Architecture is an art because it is interested not only in the original need of shelter but also in putting together spaces and materials in a meaningful manner. This occurs through formal and actual joints. The joint, that is the fertile detail, is the place where both the construction and the construing of architecture take place. (Furthermore, it is useful to complete our understanding of this essential role of the joint as the place of the process of signification to recall that the meaning of the original Indo-European root of the word *art* is "joint.") As Kahn has said,

> The joint is the beginning of ornament
> And that must be distinguished from
> decoration which is simply applied.
> Ornament is the adoration of the joint.[37]

NOTES

1. Philip Johnson, "Architectural Details, *Architectural Record* (1964), pp. 137–47.

2. W. S. Heckscher, "Petites Perceptions," *Journal of Medieval and Renaissance Studies* 4(1974), pp. 101ff. The idea of the process of signification in the details can be traced through Leibniz to Ramon Lull.

3. In jotting down those data concerning the adage, I had a *lapsus calami*, and instead of spelling God with one *o*, I spelled it with two *os*. Later on the same page of my notebook I scribbled down a note taken from a passage from Vitruvius's treatise on architecture, *De Architettura* (ed. and comm., S. Ferri; Rome, 1960), p. 10. A few days later, while reviewing those notes, I was amazed by the presence of the quasi-Platonic transcription of a quasi-Aristotelian maxim–i.e., "Good lies in the detail"–next to a note stating that Callimacus, the mythical designer of the Corinthian capitol, whose name in Greek means "He who fights strongly for beauty," had been nicknamed *Katatexitechnos* by the Athenians. By this long and complicated alias the Athenians recognized Callimacus's work as the result of an activity that proceeded with rational method toward a specific productive aim and is a knowing in the doing. *Technè* is reflection in action embodied in the details (M. Isardi Parente, *Technè*; Florence, 1966). This curious misspelling accident and association of words brought me to consider the role of *technè* in the production of architecture and in the process of architectural signification.

4. In the architectural detail, the practical norms (technology) and the aesthetic norms (semiotics) come together in a dialectical relationship. The detail is the unit of architectural production. See for the origin of this theory in the eighteenth century: Marco Frascari, "*Sortes Architectii* in the Eighteenth Century Veneto," Ph.D. diss., University of Pennsylvania, 1981.

5. For a survey and a discussion of the different elements and theories developed in architectural semiotics see: Martin Krampen, *Meaning in the Urban Environment* (London, 1979), pp. 6–91.

6. J. Labatut, "An Approach to Architectural Composition," *Modulus* 9 (1964), pp. 55–63.

7. See for a different approach, but reaching the same conclusion: Roger Scruton, *Aesthetics of Architecture* (Princeton, 1979), pp. 77ff.

8. A case is the collapse of the Marciana Library in Venice. In his first Venetian building J. Sansovino, indeed a skillful "proto," used Roman detailing (*maniera Romana*), which indeed did not work in Venice. See T. Temanza, *Vite dei piu celebri architetti e scultori Veneziani*, (Venice, 1778).

9. Functions in architecture depend on both the building itself and on who uses it or organizes its use. Custom and repeated usages are the base of functions. Architecture not only performs but also signifies its functions and can be organized in four functional horizons: the practical, the historical, the social, and the individual. For a discussion of the four functional horizons and a typology of functions see J. Mukarovsky, "The Place of the Aesthetic Function Among the Other Functions in Architecture" in *Structure, Sign and Function* (New Haven, 1978), pp. 240–43.

10. The French commercial origin of the word, which differentiates between the selling of slices of pizzas and the sale of whole ones, besides clarifying that details are parts, does not help in the understanding of the detail as joint and its nonsubordinate relationship with wholes. A better and meaningful term is the Italian, *particolari architettonici*, which is also connected with the literary theories of the 18th century, for instance, Antonio Conti's idea on *particolareggiamento*.

11. Despreaux Nicolas Boileau, *L'Art Poétique* 1 (1670; Paris, 1966), p. 158.

12. G. B. Piranesi, "Parere," (1765) in J. Wilton-Ely, ed., *The Polemi-*

cal Works (Farnborough, 1972).

13. J. Soane, *Lectures on Architecture* (London, 1929), p. 177.

14. For a discussion of the origin of the *analytique* in Lodoli's garden at S. Francesco della Vigna see Marco Frascari, "*Sortes Architectii*."

15. For this role of the analytique and the process of detailing see the discussion of Antonio Conti's theory of *particolareggiamento* in Frascari, pp. 141–150.

16. G. O. Garney, *The American Glossary of Architectural Terms* (Chicago, 1887).

17. D. Ramée, *Dictionnaire general des termes d'architecture* (Paris, 1868).

18. J. Lotman, "The Dynamic Model of a Semiotic System," *Semiotica* 21, no. 3/4 (1977), p. 194.

19. Leon Battista Alberti, *De Re Aedificatoria* (Bologna, 1782). The principle of the *nihil addi* is presented in the first book, but it is theoretically developed in the sixth and seventh books. For this new interpretation of the concept see the discussion of the role of "decoration: in the small temples: "*e'ti pare che, & vi si possa, & vi si debba aggiunere*."

20. This tripartite discussion of beauty is developed by Alberti in his seventh book (IX, 5), pp. 229–30.

21. Ludwig Wittgenstein, *Remarks on the Foundation of Mathematics* (A, II, 14, j; cf. II, 47 and V, 46) (Oxford, 1956).

22. On the use of the human body as basic design reference and generator of measures see Marco Frascari, "The Search for Measure in Architecture," to be published in *Res*.

23. For the concept of analogy in architecture see Vitruvius (ed. Ferri), pp. 50ff.

24. J. Mukařovsky, pp. 240–43.

25. H. Von Helmholtz, *Ueber Geometrie* (Darmstadt, 1968), p. 218.

26. R. Torretti, *Philosophy of Geometry* (Dordrecht, 1978), pp. 162–171.

27. Erwin Panofsky, *Gothic and Scholasticism* (New York, 1946).

28. Walter Benjamin, *Illuminations* (New York, 1968), p. 242.

29. R. Torretti, p. 168.

30. Accademia Olimpica, *Carlo Scarpa* (Vicenza, 1974), p. 1.

31. J. Mukarovsky, pp. 240–43.

32. Carlo Scarpa, "Frammenti, 1926–78," *Rassegna* 7 (1981), p. 82.

33. H. Focillon, *Piranesi* (Bologna, 1962), p. 166.

34. Carlo Scarpa, pp. 83–84.

35. Carlo Scarpa, pp. 83–84.

36. For a discussion of the use of "fertile details" see the analysis of the "fertile motif" in Anton Ehrenzweig, *The Hidden Order of Art* (London, 1962).

37. Louis Kahn, *Light is the Theme* (Fort Worth, 1975), p. 43.

BIBLIOGRAPHY

Accademia Olimpica. *Carlo Scarpa*. Vicenza, 1974.

Alberti, Leon Battista. *De Re Aedificatoria* (1485). Milan, 1966.

Benjamin, Walter. *Illuminations*. New York, 1968.

Blomfield, Reginald. *The Mistress Art*. London, 1908.

Boileau, Despreaux Nicolas. *Art poétique*, I (1670). Paris, 1966.

Brusatin, Manlio. "Carlo Scarpa." *Controspazio* 3/4 (1972), pp. 2–85.

"Details." *Construction Details* (January 1914), p. 1.

Ehrenzweig, Anton. *The Hidden Order of Art*. London, 1962.

Focillon, H. *Piranesi*. Bologna, 1967.

Frascari, Marco. "*Sortes Architectii* in the Eighteenth Century Veneto." Ph.D. diss., University of Pennsylvania, 1981.

———. "The True and the Appearance; The Italian Facadism and Carlo Scarpa." *Daedalus* 6 (December 1982), pp. 37–46.

Garvey, G. O. *The American Glossary of Architectural Terms*. Chicago, 1887.

Helmholtz, H. Von. *Ueber Geometrie*. Darmstadt, 1968.

Heckscher, W. "Petites Perceptions." *Journal of Medieval and Renaissance Studies 4* (1974), pp. 100–142.

Johnson, Philip. "Architectural Details." *Architectural Record* (April 1964), pp. 137–47.

Kahn, Louis. *Light is the Theme*. Fort Worth, 1975.

Krampen, M. *Meaning in the Urban Environment*. London, 1979.

Labatut, Jean. "An Approach to Architectural Composition." *Modulus* 9 (1964), pp. 55–63.

Lotman, J. M. "The Dynamic Model of a Semiotic System." *Semiotica* 21, no. 3/4 (1977), pp. 193–210.

Mukařovsky, Jan. *Structure, Sign and Function*. New Haven, 1978.

Panofsky, Erwin. *Gothic and Scholasticism*. New York, 1946.

Parente, M. Isardi. *Technè*. Florence, 1966.

Piranesi, G. B. *Polemical Works*. J. Wilton-Ely, ed. Farnborough, 1972.

Ramée, Daniel. *Dictionnaire general des termes d'architecture*. Paris, 1868.

Scarpa, Carlo. "Frammenti, 1926–78." *Rassegna* 7 (1981).

Scruton, R. *Aesthetics of Architecture*. Princeton, 1979.

Serlio, S. *Trattato di architettura*. 1619.

Soane, J. *Lectures on Architecture*. London, 1929.

Torretti, R. *Philosophy of Geometry from Riesmann to Poincaré*. London, 1978.

Van Pelt, J. V. "Architectural Detail." *Pencil Point* (May 1921, June 1921).

Vitruvius. *De Architettura 7 vols*. S. Ferri, ed. and comm. Rome, 1960.

Wittgenstein, Ludwig. *Remarks on the Foundation of Mathematics*. Oxford, 1956.

1. *Music Center under construction.*

BUILDING ORDER

Herman Hertzberger

IN SIMPLE TERMS, you could say that building order is the unity that arises in a building when the parts taken together determine the whole and, conversely, when the separate parts derive from that whole in an equally logical way. The unity resulting from design that consistently employs this reciprocity may in a sense be regarded as a structure. The material (the information) is chosen on purpose, adapted to suit the requirements of the task in question, and, in principle, the solutions of the various design situations are permutations of or at least directly derived from one another.

As a result there will be a distinct, one could say family relationship between the various parts. (Following this train of thought, one sees that there is an obvious comparison with that outstanding example, of structure, the genetic system.) Of course, every well-designed building has a consistent idea with a distinct thematic unity behind it, a unity of vocabulary, material, and building method. But here the essential thing is design based on a consistent strategy, whereby the aim is a harmony of parts and spaces capable of accommodating a variety of functions and activities. This train of thought, inspired by structuralism, which superficially resembles classicism,[1] attempts to square accounts with the somewhat contradictory functionalistic slaving to find a specific form and a specific spatial organization for every different "function."

Design that seeks the largest common denominator, the "set" of all the requirements under discussion in a particular task (i.e., the program in its widest sense), employs a different strategy and demands a fundamentally different outlook from the architect.

• • • • •

The elements of the Vredenburg Music Center in Utrecht, the columns in particular, may further illustrate my idea of building order and how I arrive at it. Seen from the outside, the complex appears to have an arbitrary form and does not exactly give the impression one would expect of an autonomous building.

The Music Center may be regarded as a transition between the old town center of Utrecht and the new shopping center "Hoog Catharijne," which is too large and has penetrated too far into the city.

The building as a whole is articulated in a number of parts that are separated from each other, as if they were blocks in the city, by glass-covered shopping arcades. These are on the same scale as the alleys in the town center and are connected with the central area of the shopping center. The Music Center itself is surrounded by shops and other facilities located in the arcades, such as the information center, offices, and a couple of restaurants.

Not only does this organization create a link with the area outside the Music Center, but it also takes away the autonomy of the building. An autonomous, detached, and more or less monumental building would have been less in keeping with the intended emphasis on general accessibility.

When one thinks of a theater or concert hall, images of detached monumental buildings, directly or indirectly associated with temples, come to mind. The nineteenth century was so prolific in this type of building that later examples, though perhaps less literal regarding form language, were unable to break away from it.

The plan of the Vredenburg Music Center departed from this tradition and is intended to be an

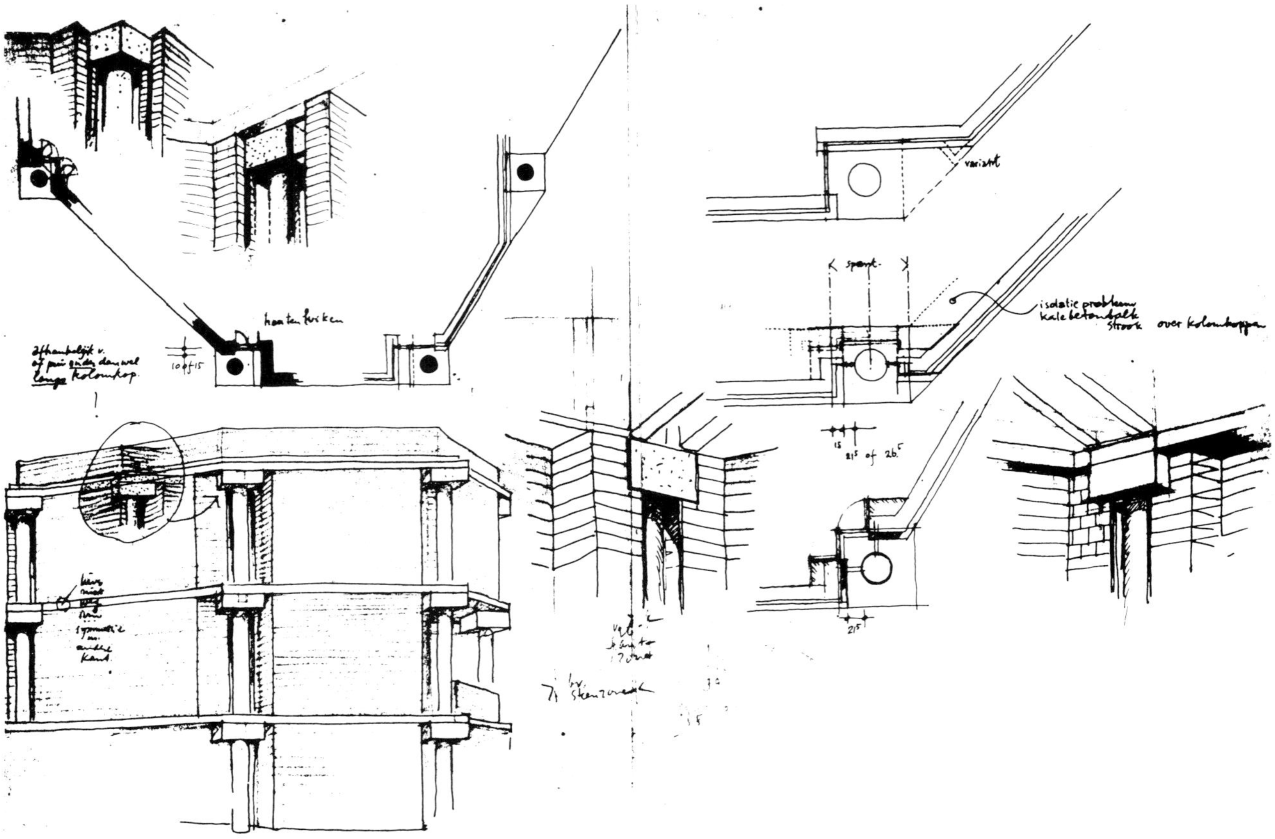

2. *Sketches.*

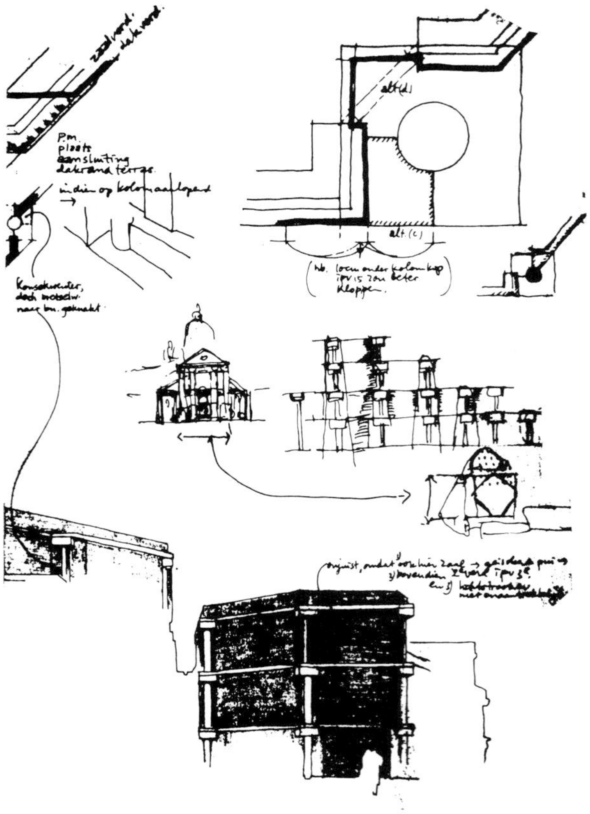

3. *Sketches.*

attempt to arrive at a less formal, less stately, and perhaps, therefore, more accessible form. A breakthrough has been made regarding not only the concept but also the "mechanism" of accessibility. One does not enter the building in an obvious way through a main entrance, but rather in stages. First in the arcade, you pass (or go), as if into a shop, through one of its many entrances into the lobbies of the Music Center itself.

These same ideas guided the design of the periphery and its composition as a multitude of facets. And because these facets are all constructed of the same material, they are in fact many views of the same thing. In other words, the emphasis is more on making the parts interpretable than on the whole, yet the whole is represented in the parts.

This approach enables you to look at the whole from many different sides, and the attention becomes relativized. The building elements become more autonomous, emancipated as it were, and the different ways in which they are "mounted" in the many corners bring them into changing relationships with one another, depending on the place.

All the same, despite its capricious shape, the unity of material and building elements and the way in which the latter are put together enable the building to

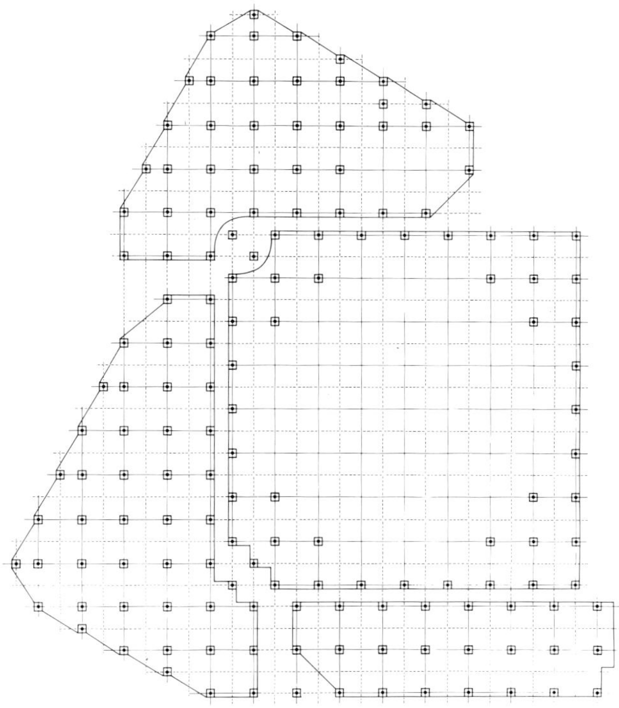

4. Plan of the column grid.

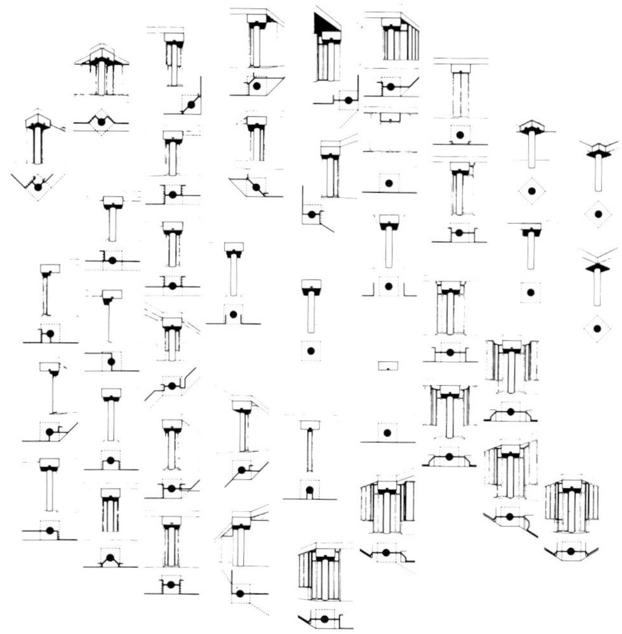

5. Various column conditions.

be clearly interpreted as one architectural language. This has been accomplished consistently throughout the building, its manifestation both inside and outside being the same in principle. Only inside is there also the lining, consisting for the most part of wood. Use of the same material both inside and outside the building creates a relativization of inside and outside and is intended to contribute to the expression of accessibility.

The columns that recur everywhere, with their distinct and easily recognizable form, play an important part in the building order. Standing in rows on a square grid, they recur in every space at an equal distance, demarcating equal areas throughout the building.

The column system forms a minimum system of arrangement that allows great freedom when it comes to filling it in, and that is able to coordinate the large variety of parts issuing from the highly complex program. While ensuring the unity of the whole, the column system is an encouragement to shape every spot according to the requirements of each particular situation.

It is easier, or at any rate simpler, to connect rectangular columns with walls, or to place furniture against them. This potential was already made use of, in fact, in such earlier buildings as the Centraal Be-

6. Exterior view.

7. Exterior view.

8. Exterior columns.

9. Exterior columns.

10. *Exterior columns.*

11. *An exterior column with a balcony and bench.*

12. *An exterior column.*

13. Coffee area at the Centraal Beheer Office Building.

14. View of Die Drie Hoven interior with square columns that accommodate furniture placed against them or walls connecting into them.

15. Restaurant interior at the Music Center with boxed columns.

heer office (Fig. 13) and the old people's center De Drie Hoven (Fig. 14). However, in the case of the Music Center, which may be characterized primarily as a series of extensive, continuous open spaces, it was more natural to make as many columns as possible freestanding.

When it comes to free-standing columns, round ones are certainly the most beautiful, if only because their presence among people is so benign and gentle. Otherwise, where it was necessary to push tables against them, as in a restaurant, the owner solved the problem in his own way, thus relentlessly proving the theory (Fig. 15).

Always standing "in the way" but never obstacles, round columns have a strong personality that is stressed in this building by their square heads: an overstatement of the form necessitated by the construction. The most important function of aligning these heads all in one direction, is to make possible the junction of the ceilings of various heights coming to them from different directions. Further, their wider heads keep the adjoining partitions at a distance, so helping to create a separate free zone around the columns. Where columns stand in partitions and exterior walls, the walls remain at a lesser or greater distance depending on the amount of glass required in a particular place. Openings are mainly to be found in this zone around the columns; seldom are they holes in a wall.

The way in which the columns are positioned, free-standing where possible in the free zone enveloping them, creates a motif, repeated throughout the building in many forms, that produces a recognizable and distinctive image. The column is designed so as to make different spatial experiences possible in each particular place, whereby its intrinsic characteristics remain the same even though it functions in different situations. Its guise varies according to the amount of openness or seclusion required: you could say it wears a different costume for each role. In this way the column, while determining the image of a place, is itself determined as an image at the same time. The column system generates freedom: it has the "competence" to inspire the appropriate "performance" for each situation, thus being an instrument able to create a distinct building order despite the absence of recurrent spaces.

The building order of a project is the result of a precise visualization of all that is going to be done with it, and also of how it might be used at a later stage. It anticipates the *performances* most likely to occur. From this a *competence* is constructed by a process of induction. In fact, every building task is an incentive to develop an individual arrangement, one that arises from the specific character of each task. Since each task represents a certain kind of mechanism, they are not interchangeable. In each case different requirements are emphasized, but where structure is concerned it is always a question of the paradox of an order that generates freedom: *a horizon in your plan.*

16. View of the Music Center lobby.

NOTE

1. The order of a building designed according to a consistent idea with a clearly legible theme, which creates unity of vocabulary, material, and style as well as spatial form, inevitably seems to be related to classicism.

Classicism does indeed work with the same elements; however, they are only of formal significance, just as the stereotyped blocks of a construction kit determine the structure of all buildings made with it without being able to express differences in use and function.

At present you see classicism, persistent as it is, being dug out again and rationalized as respect for the past, not to mention the feeling of familiarity that recognition supposedly creates. But we must realize that what is being recognized are the authoritarian relationships that always accompanied classicism. Whether they expressed the hegemony of the bourgeoisie or of the workers, it is always a question of the expression of power and, thereby, reinforcement of that power.

Although in principle the vocabulary of classicism can be used to express other than authoritarian relationships (and of course there are plenty of historical examples), it is not really in its nature to do this. Every language has its own particular range of expression and limitations. That is to say one language cannot express everything, but one is better able to express a certain range of experiences and emotions than another.

Classicism implies formal organization, whereby axes and symmetry are used to compel the attention to go in a certain direction.

Classicism can only be used to express the classical mechanism, that is to say, a hierarchical organization that always implies a dominant factor on which the entire system is based, which all too soon leads to univalency and the tendency to impose.

So this compelling of attention, inherent in the mechanism of classicism, is what makes it go hand in hand with the expression of authoritarian relationships, and therein lies its danger.

If there is one thing that the twentieth century has produced it is the awareness of equivalence and consequently the breakdown of hierarchical systems. This awareness results from the discovery of the enormous variety of cultures "dug up" in the twentieth century and from the "musée imaginaire" of forms of expression thus yielded. In this light eclecticism and copies of form reverting to a limited portion of "familiar" past are just not feasible any more.

Now, after this essential expansion of our awareness, we have no choice but to make forms containing much of what all people, primitives and moderns alike, carry in their heads as archetypes, forms that resonate with these archetypes and thus represent the collective unconscious. Our columns and beams, then, ought to have a polyvalence capable of evoking many associations, so that you keep thinking you've seen them before, *although that is not the case.*

17. Sketches of the interior of the Music Center.

1. Frontispiece from The Seven Lamps of Architecture, *John Ruskin, 1849.*

ON TRUTH AND LIE IN ARCHITECTURE

Karsten Harries

I

RUSKIN took "the glistening and softly spoken lie" to be the greatest source of mischief in the world.[1] Truth seemed to him like a home that, willfully forsaken, continues to call. To this home Ruskin wanted to recall his contemporaries, and to this home he wanted to lead back especially an architecture that he thought had succumbed to the seductive power of the lie all too willingly.

> From that one surrender of its integrity, from that one endeavor to assume the semblance of what it was not, arose the multitudinous forms of disease and decrepitude, which rotted away the pillars of its supremacy. . . . It was not the robber, not the fanatic, not the blasphemer, who sealed the destruction they had wrought; the war, the wrath, the terror, might have worked their worst, and the strong walls would have risen, and the slight pillars would have started again, from under the hand of the destroyer. But they could not rise out of the ruins of their own violated truth.[2]

Today we may find it difficult to take seriously the overcharged rhetoric of *The Seven Lamps of Architecture*. What does Ruskin mean when he speaks of architecture's "violated truth"? Most often we attribute truth to assertions. An assertion is true when it accords with what is the case. But is architecture sufficiently like language to allow us to liken buildings to assertions that may be judged true or false?

One obstacle to such talk is our inclusion of architecture among the visual arts. The demand that a work of art be a self-sufficient whole has become commonplace. This requirement suggests that only as long as a building is governed by its own aesthetic perfection can it be considered a work of art, i.e., architecture. The point of such a work is not to refer beyond itself to some other reality that it represents or is about. Truth of representation has nothing to do with artistic success. It would seem that talk of truth and falsity has no more place in architectural theory than it does in the theory of art.

To be sure, we may call a work of art "true" and mean by this that such a work is true to itself, that it coheres and by its coherence convinces us that it has to be just as it is, that it could not be otherwise without weakening or destroying its aesthetic integrity. To be true in this aesthetic sense a work must be strong enough to so absorb our attention that we are not tempted to judge it by how ill or well it conforms to some paradigm or type or to the rules of some style. Strength of self-reference and coherence should discourage attempts to refer such a work to or measure it by anything outside itself.

I have no objection to such talk. It helps to remind us that when we call something "true," we often mean that we experience it as being as it should be. I doubt, however, that an understanding of truth as aesthetic coherence does justice to what we usually mean by the term, and I wonder whether it captures what is meant by those who praise a building for its truth or condemn it for its lies. Ruskin seems to me right to link truth to assertion. Only if architecture, too, represents something, can it also misrepresent it, that is, lie. But what do works of architecture assert or represent?

II

Some structures are representations in the most obvious sense. Robert Venturi has called such buildings

2. *Choir, Beauvais cathedral, 1225–1568.*

3. *"The Long Island Duckling," from* God's Own Junkyard.

"ducks," in honor of a duck-shaped drive-in on Long Island.[3] But the very fact that in such cases the representational function of the building cannot be overlooked makes us hesitate to speak here of architecture. We certainly would not want to measure a building's success as architecture by how ill or well it represented a Long Island duck. When we speak of truth in architecture we are not thinking of adequate representation in this sense.

Often we call something "true" to suggest that it is in accord, not with some fact, but with its essence. By a "true hammer," we mean thus a hammer that functions as a hammer should function. True architecture would then be architecture that agrees with the essence, or with the rules, or with the grammar of architecture.

But does such a view enable us to make sense of architectural deceits or lies? To call a building a lie is to claim not just that it is unsuccessful because it fails to do justice to the essence of architecture (whatever that may be), but that it masks this failure in some way, that it misrepresents itself, and that such misrepresentation is deliberate, that the building is meant to deceive. But this returns us to the problematic claim that architecture can possess a representational function.

I shall claim that it is indeed its power of representation that distinguishes architecture from mere building, that true architecture does not just agree with the essence of architecture but represents or asserts such agreement.

How then are we to judge the historicist architecture for which Ruskin pleads? Can we claim truth for structures that, built in obedience to the "grammatical laws" of Gothic or classical architecture, also proclaim such obedience? Does such obedience mean obedience to the essence of architecture? If so, Yale's neo-Gothic Harkness Tower, built in 1921 with medieval methods of construction, would have to be considered a paradigm of truth in architecture. Few would want to make such a claim. Ruskin's "grammatical laws" would appear to be no more than conventions that, divorced from their historical context, no longer carry conviction. Must not submission to such conventions lead to a theatrical architecture, an architecture that lies, even if a building's appearance in no way masks its structure? Ruskin, of course, would not have agreed.

> It is almost impossible for us to conceive, in our present state of doubt and ignorance, the sudden dawn of intelligence and fancy, the rapidly increasing sense of power and facility, and in its *proper sense*, of Freedom, which such wholesome restraint would instantly cause throughout the whole circle of the arts. Freedom from the agitation and embarrassment of that liberty of choice which is the cause of half the discomforts of the world; freedom from the accompanying necessity of studying all

4. Harkness Tower, Yale University, James Gamble Rogers, 1921.

5. Houses of Parliament, London, Charles Barry and A. W. N. Pugin, 1837–60.

> past, present, or even possible styles; and enabled, by concentration of individual, and co-operation of multitudinous energy, to penetrate into the uttermost secrets of the adopted style, the architect would find his whole understanding enlarged, his practical knowledge certain and ready to hand, and his imagination playful and vigorous, as a child's would be within a walled garden, who would sit down and shudder if he were left free in a fenceless plain.[4]

In *The Seven Lamps of Architecture* insistence on freedom mingles with a fear of freedom, insistence on truth with a fear of truth, leading to a perversion of truth. (By now we should have learned to suspect all dreams of walled gardens in which we are allowed to play, innocently, as children play.) Freedom, Ruskin insists, should not be confused with "liberty of choice"; rather, it requires what he calls "wholesome restraint." This distinction between a negative freedom that degenerates into merely subjective, arbitrary choice that lets us lose sight of truth and a positive freedom that means submission to the truth recalls Rousseau, for whom freedom meant submission to the general will, or Kant, who tied freedom to obedience to that universal law that reason gives to every human being. Obedience here means autonomy, submission to man's own essence. But does reason provide architects with such a law? Does architecture know something like its own categorical imperative? If so, one might try to show that Gothic architecture is obedient to architecture's truth.

Ruskin settles for much less. "Wholesome restraint," as he understands it, is provided by the "grammatical laws" of an adopted style. *Which* style is adopted does not seem to matter all that much. Rather, what matters is that a style be such that a community can find in it something like a common architectural language, a binding code, and that architects will be able to work once more subject to a ruling architectural convention. The rule of this convention is to liberate architects from the confusions of an architecture open to countless past and possible conventions and, for just that reason, governed by arbitrary, subjective whim. A book like William Hubbard's *Complicity and Conviction* demonstrates the continuing appeal of such suggestions as it betrays the underlying despair.[5]

While Ruskin does not claim that the essence of architecture finds its uniquely adequate realization in the classical or Gothic styles, he does give reasons for his preference for these particular styles. He points to their "facility of adaptation to general purposes" and to the sympathy with which they "would be popularly regarded."[6] Such sympathy could indeed be expected from Victorian gentlemen who liked to see themselves in the image of the heroes of antiquity or of medieval knights. But must not submission to popular taste pervert freedom? And perverting freedom, must it not also pervert truth? Small wonder that the historicist architecture Ruskin invites was soon condemned for its dishonesty, for its willingness to substitute theatrical appearance for reality. To be sure, we have grown more receptive to the charms of the theatrical architecture of the nineteenth century; we have grown to appreciate its false Gothic, its false Renaissance, its false baroque. But would we claim truth for such architecture? If so, what would we mean by "truth"?

Postmodernists have found it difficult to dismiss Nietzsche's suggestion that "to be truthful means using the customary metaphors–in moral terms: the obligation to lie according to a fixed convention, to lie herd-like in a style obligatory to all."[7] Applied to architecture, this statement would mean: to be truthful is to build using an accepted vocabulary in accepted ways. Truth in architecture is reduced to obedience to a dominant style. Architecture violates that truth when individual architects refuse the authority of such a supra-individual style.

But did the nineteenth century, do we today, know architectural truth even in this relative sense? Ever since the death of the rococo and the birth of neoclassicism, it has become difficult to point to an architectural style that is felt to be obligatory. It was precisely the weakening of the hold of the architectural system established by the Renaissance that raised questions about the essence of architecture and led to demands for truth in architecture.

Once such demands have been made, submission to an adopted style, such as the submission Ruskin envisions, must be considered a surrender to a code that has long since lost its binding force. We may well grant Ruskin that the loss of the paradise of style is inseparable from that liberty of choice that is, he suggests, "the cause of half the disorders of the world." We have become too reflective, too critical, too free, to deceive ourselves into thinking mere conventions a source of obligations; we have learned to suspect invocations of "the truth." All too often they are instruments of deception, especially of self-deception.

Ruskin's use of the term is a case in point. To call something "true" is to suggest deliverance from the vicissitudes and distortions of time and place. Truth is not bound to a particular perspective. Thus it escapes the distortions of a particular time and place. But Ruskin's historicism locates truth in what he knows is merely a historical convention, endowing what is relative with a dignity that belongs only to what is absolute. Postmodernism's ironic play with historicist architecture helps to rid the latter of its claim to truth and thus of its lies. But the power of irony is the power of negation; to negate a lie is not to put a truth in its place. Hence, when postmodern architects play with past building conventions, such play may liberate by revealing those conventions' merely conventional character, but it does not lead beyond inherited conventions to architecture's truth.

III

Buildings speak to us of their use. The most obvious task of architecture is to help human beings establish control over their environment: to provide shelter, to establish frames for different activities, to grant a sense of being at home in the world. This task suggests that we may speak of truth in architecture when buildings serve and reveal their function.

Most buildings would seem to do this: most churches look like churches, most houses look like houses, most office buildings look like office buildings. Similarly windows not only are, but look like windows, doors not only are, but look like doors, roofs not only are, but look like roofs, while our expectations of what windows, doors, and roofs should look like will depend on our experiences with buildings. These features will, therefore, reflect variations in landscape and history.

It is not surprising that the Enlightenment appealed to function when it sought to measure and reform long-established architectural practice by what it took to be the essence of architecture legitimated by reason and nature. Carlo Lodoli, one of the Enlightenment's leading theorists, insisted therefore that "In architecture only that shall show that has a definite function, and which derives from the strictest necessity."[8] The architectural manifestoes of the twentieth century demonstrate the extent to which their authors remain heirs of the Enlightenment.

When taken seriously, more seriously than Lodoli himself was willing to take it, his commandment leads not only to a critique of ornament, which, by its very nature, is something added on to what necessity demands, but to a critique of the very conception of architecture as an art. A rigorous functionalist will reject all attempts to have a building display its obedience to the demands of function. For to say of an object that it is functional is not at all to say that it was designed to assert its functionality. Take a simple tool—a hammer, for example. There are of course better and worse hammers. And we can think of hammers designed to have the look of good hammers—that appearance may well help to sell them. But concern for appearance is not part of a strictly functional approach; it does not help produce better hammers. The result of such concern may indeed be just the opposite.

Similarly, there are countless modern buildings designed to have the look of being functional, often at considerable cost, that is to say, at the price of true functionality. We can thus distinguish rigorous functionalism from a rhetorical aesthetic functionalism. The latter aims not so much at functional buildings as at representations of functional buildings. Think of the Seagram Building or of the Centre Pompidou. Aesthetic functionalism seeks to give buildings the look of embracing science and technology, a look communicating an ethos and calling us to a way of life.

A rigorous functionalist will deplore such concern with appearance. He has to condemn all architecture not content to be a branch of engineering. Buildings should be machines. To be sure, machines will often reveal their function. But this disclosure should not be the point of their design. What matters is that they function, not that they make assertions about their functioning.

There are, of course, machines designed to make such assertions. Think of cars, or of toasters, or of telephones; examples are legion. Someone concerned strictly with function has to consider this aspect of their design superfluous ornament. He might cite Ruskin, who thought it just common sense "not to decorate things belonging to purposes of active and occupied life. Wherever you can rest, there decorate;

where rest is forbidden, so is beauty. You must not mix ornament with business, any more than you may mix play."[9]

Presupposed by such a functionalism is that compartmentalization of life that finds expression in slogans like "art for art's sake," "business is business." Ruskin, of course, would not have us extend this functional approach to all building. He would not have us understand houses or churches as machines meeting the purposes of "active and occupied life." Precisely for this reason they become the proper province of the architect and of ornament. Architecture, from this view, is architecture because it goes beyond, indeed violates, truth to function, because, in this sense at least, it lies.

But perhaps we should question the meaning that here is given to "function" and "lie." By now we have had to learn that even the worker in an office or on an assembly line cannot be reduced to a mere worker. "Rest" and "active and occupied life" may not be separated very sharply. When attempts have been made to insist on that separation, they have resulted not in greater, but in less, efficiency. Human beings are not as easily compartmentalized as technological thinking would have them. The lesson to be learned is that not only houses and churches, but office buildings and factories, too, should be more than machines.

Corbusier knew that as long as the problem of the house had not yet been stated, his programmatic definition of the house as "a machine for living in" could only have a utopian ring. But the problem of the house will never be stated with sufficient precision to allow an engineer to provide the solution, for the simple reason that architecture, beyond providing physical frames for human activities, also interprets to human beings their place in nature and society. This place is not given as an already established matter of fact. As the natural and social environment changes, as hopes and dreams change, this place needs to be established ever again. Architecture is true to its own essence only when it contributes to such establishment.

What I have called, somewhat misleadingly as it turns out, rigorous functionalism, makes no such contribution. Nor does it call for architects. Aesthetic functionalism, on the other hand, by its representations of a functional architecture, addresses the demand that architecture interpret our place in a very direct way, although we may well wonder whether the place to which it calls us does not do violence to human existence even as it claims to do justice to the shape of modernity.

IV

If buildings speak to us of their use, they also speak to us of the materials of which they appear to be made. Again it is Lodoli to whom we owe the commandment that architecture "conform to the nature of materials."[10] The architectural lie has often been under-

6. Branch bank, Washington, D.C., Martin and Jones, 1983.

7. Seagram Building, New York, Mies van der Rohe, 1956–58.

8. Die Wies, Interior, Bavaria, Dominikus Zimmerman, 1754.

9. "Re-presentation of material."

stood, in Ruskin's phrase, as "falsity of assertion respecting the use of materials."[11] Ruskin therefore attributed the decline of Gothic architecture to its willingness to sacrifice the quality of stone to forms appropriate to wooden construction. Often some cheaper or more easily worked or maintained material substitutes for one that would involve higher costs, while efforts are made to conceal the substitution (e.g., aluminum siding, or wooden houses with brick veneers). The history of architecture is full of such "lies."

Consider a church like Dominikus Zimmermann's pilgrimage church Die Wies, one of the acknowledged masterpieces of the rococo. The church's frescoed vault appears like a weightless sail. The impression of weightlessness is enhanced by delicate supports that seem incapable of carrying a substantial load. To create this impression, the architect takes advantage of our usual experience with buildings, which leads us to assume such vaults to be made of stone. This vault, however, is constructed of timber, lath, and plaster. Because of our expectations derived from masonry vaults, the vault of Die Wies has to appear magically light. For the sake of that appearance, which is to enhance the sacred quality of the space, the architect does not hesitate to make use of what Ruskin might well consider an unpardonable lie. And Ruskin would not be alone with his condemnation. Schopenhauer, to give another example, claims that

> It is absolutely necessary for an understanding and aesthetic enjoyment of a work of architecture to have direct knowledge through perception of its matter as regards its weight, rigidity, and cohesion. . . . if we were told clearly that the building, the sight of which pleased us, consisted of entirely different materials of very unequal weight and consistency, but not distinguishable to the eye, the whole building would be as incapable of affording us pleasure as would a poem in an unknown language.[12]

Architecture, Schopenhauer insists, should be written in a language known to us. This language cannot be reduced to a language of beautiful forms; beyond form and symmetry, it is the "conflict between rigidity and gravity that constitutes the life of the stone." False understanding of a building's materials leads to a false understanding of this conflict; as a result, the building will possess only an apparent, not a genuine life. Similar consideration leads Ruskin to claim "that building will generally be the noblest, which to an intelligent eye discovers the great secrets of its structure, as an animal form does, although from a careless observer they may be concealed."[13]

To insist that a building display its materials and structure is to insist on more than the kind of regard for the different properties of different materials forced on any builder by his task. Vitruvius has emphasized the importance of such understanding. But while every craftsman must understand the materials of his trade if his work is to succeed, such success is hardly a good reason for calling the product of his labors "true." Just as regard for function does not mean that an attempt has been made to display that function, so regard for materials does not mean that the object made either makes an assertion about its materials or calls their nature to our attention–quite the opposite. As Martin Heidegger points out, in making equipment, material is not only used, but used up. "It disappears into usefulness. The material is all the better and more suitable the less it resists perishing in the equipmental being of the equipment."[14]

When tools work well, we often forget what they are made of. Something like that is also true of many buildings. But we may want to insist that a building that is also a work of art, i.e., architecture, lets us regard materials in a very different way. As Heidegger says of a Greek temple, such a building

> . . . does not cause the material to disappear, but rather causes it to come forth for the very first time and to come into the Open of the work's world. The rock comes to bear and rest and so first becomes rock; metals come to glitter and shimmer, colors to glow . . . .[15]

We can thus distinguish buildings that merely presuppose regard for materials from architecture that reveals the essence of these materials. Such revelation presupposes that material is worked in a way that invites us to step back from our everyday involvement with things. Usually we are too busy, too caught up in our activities, to really look at the stuff of which the things around us are made. The work of art, more especially the architectural work of art, tears away the veil familiarity has cast over things. Its use of materials is at the same time a re-presentation of these materials in the most literal sense. Such re-presentation makes visible their usually overlooked and taken-for-granted essence. It shows us what these materials are in truth.

But even if architecture can be re-presentational in this sense, does truth to materials provide the architect with something like an imperative? One might think of a similar imperative in painting, which would have the painter re-present the materials of his art so that paint reveals itself as paint, canvas as canvas. It is not difficult to point to works of art that have taken this commandment to heart, that for the sake of such re-presentation have forsaken the traditional representational function of art. But there is also a great deal of art that for the sake of representation invites us to forget its materials in much the same way in which a church like Die Wies lets us forget its materials. Should we speak in such cases of art having succumbed to the lie?

Jacques Maritain, in fact, makes this charge, inviting comparison with John Ruskin's condemnation of lies in architecture:

> With the sixteenth century the lie installed itself in painting, which began to love science for its own sake, endeavoring to give the *illusion* of nature and to make us believe that in the presence of a painting we are in the presence of the scene or subject painted, not in the presence of a painting.[16]

Like Ruskin, Maritain associates truth with medieval, lie with Renaissance and post-Renaissance art. To escape from the lie, painting has to present itself to us as such, as painting, has to assert its very material, its painterly essence. Similarly, we may understand truthful architecture as architecture that asserts itself as architecture, as a very material bounding of space.

There are ready objections to such suggestions. A modern aesthete might insist that art is for art's sake, not for the sake of truth. It is interesting to note that Francesco Algarotti, to whom we owe the report of Lodoli's commandment that architecture conform to the nature of its materials, was himself too much a child of the rococo, or perhaps too much of an aesthete, not to fear the consequences such a commandment might have. Challenging Lodoli, he insisted that illusion is more beautiful than truth. More than a century later, Nietzsche was to write that "We possess *art* lest we *perish of the truth*."[17] According to Nietzsche, we need the lies of art to transfigure and to interpret reality. Are not medieval cathedrals and rococo churches lies in this sense?

The question invites reconsideration of the opposition between the lie of postmedieval art and the truth of medieval art that Maritain and Ruskin assert. Ruskin himself observes that in Gothic architecture, too, "a certain deception" occurs. Here, too, the visible supports often seem insufficient to carry the load of the vault, which, like the vault of Die Wies, is rendered unnaturally light. Ruskin excuses this effect by pointing out that we are not really deceived. Nor is there an attempt at deception. The tension between what we know to be the case and the contrary appearance becomes here a "legitimate appeal to the imagination."[18] Such tension gives architecture something like the power of a metaphor. The building presents itself to us as other than we know it to be. What we know to have weight appears weightless; what we know to have matter and substance seems immaterial, spiritual. Medieval builders sought such appearance not to have us mistake it for reality, but to help establish the church building as a figural representation of the ideal Church. The same understanding still guided those who built Die Wies. For the sake of such representation they were not afraid to create a theatrical architecture that in many ways violated the truth of materials.

But why should truth to materials matter at all? Why should we not allow the rococo the "lies" of its false vaults and scagliola columns? Presupposed by the Enlightenment's condemnation of such "lies" is the association of such architecture with what was

10. Renaissance Center, Detroit, John Portman.

considered a theatrical, false way of life, to which was opposed the ideal of a simpler life, governed by reason and obedient to the demands of nature. Like truth to function, truth to materials, or rather, the aesthetic appearance of such truth, speaks to us of such a life. It, too, has a rhetorical function. It calls us back to nature, back to the earth.

This rhetorical function becomes particularly apparent when we consider a variety of materials, especially materials made available only by modern technology, such as steel, concrete, plastics, and tinted glass. It is easy to understand, if difficult to endorse, Ruskin's assumption that "true architecture does not admit iron as a constructive material."[19] Ruskin knew about the cultural relativity of his judgment. He even ventured the prediction that "the time is probably near when a new system of architectural laws will be developed, adapted entirely to metallic construction."[20] Before it could yield a new understanding of what constitutes truth in architecture, such a system would have to generate a new architectural tradition, a new style. By now we have become so used to new building materials that a tower built according to all the rules of Gothic architecture must strike us as theatrical, as false, not to its materials, but to the spirit of the age. Ruskin's "truth" has become our "lie."

Beyond such relativity, different building materials do carry different meanings. Structures of steel and concrete communicate a very different ethos than do structures of brick and wood. Different materials are differently affected by the passage of time and speak of different attitudes to time. Truth to materials made available only by our technology suggests not so much obedience to nature as a will to power that wants nature overpowered and transformed. There are, no doubt, many who would still call such "truth" a lie, born of self-deception; presupposed would be a very different ideal of how human beings ought to live. Truth to that ideal would demand resistance to the rule of technology.

V

All architecture that invites talk of truth to function or to materials succeeds as architecture only because it presents the appearance of such truth. This appearance has a rhetorical function: it claims obedience to the essence of architecture and offers a determination of that essence.

As the word "appearance" suggests, such presentation cannot be defended by appeals to truth to function or materials. What distinguishes architecture from mere building is precisely its refusal to obey Lodoli's commandments that in architecture only that shall show which has a definite function and that architecture be true to materials. To Lodoli's commandments we can oppose Goethe's observation that building becomes architecture only when it represents itself.[21] A house should therefore not only be a house, but represent a house, a church not only be a church, but represent a church, a factory not only be a factory but represent a factory. Architecture represents architecture. Representing architecture, it cannot help but communicate a particular understanding of what matters in architecture, a particular ideal of building, and thus of dwelling.

How does architecture represent architecture? Goethe points to a Greek temple's translation of wooden columns into columns of stone. A familiar architectural element is translated into a different medium. But translation need not be that literal. Some dislocation is inseparable from our experience of even the most functional building as an aesthetic object. A window, for example, that is experienced as part of an artistic whole no longer merely functions as a window, but also represents a window. Buildings that are also works of art cannot but represent architecture.

If the aesthetic quality of an object dislocates us and opens our eyes, the simple dislocation of familiar objects invites an aesthetic response. Postmodern architects have delighted in exploiting this possibility. Michael Graves's use of the keystone motif in the Plocek House provides a much discussed example.

Long before Graves the keystone had lost its original structural significance and had been translated into an ornamental motif. As such it had become part of the architectural vernacular. Graves translates this translation and thereby opens our eyes to it.

The poetry of architecture is inseparable from such dislocations. And just as someone who insists on literal speech can easily show that a poet's metaphors are false, so an architectural literalist, i.e., someone who insists on structural simplicity and honest use of materials, will have little difficulty pointing out "architectural lies" in all architecture worthy of that name. Think of Gothic cathedrals or of rococo churches like Die Wies or even of the seemingly much more literalist architecture of a Gropius or a Mies van der Rohe. Consider once more the Seagram Building: it appears so simple, so reasonable. But it is precisely this appearance that would seem to have mattered to the architect. In this case, too, we have to say that architecture succeeds as architecture not despite its "lies," but because of them. A certain violation of truth to function and materials is a condition of architecture. Only such violation allows a building not just to be, but also to mean, to represent architecture. But instead of speaking here of "lies," it would be better to give truth a different meaning. Truth in architecture is truth of representation.

We have become so used to divorcing the essence of art from representation and, even more to opposing architecture as the essentially abstract art of bounding space to all representational art that talk of representation may well seem to miss what matters in architecture. The example Venturi provides with his Long Island duck can only reinforce such misgivings.

Yet we should not forget that much of the architecture of the past was representational. Houses and temples were built to repeat and represent real or ideal paradigms sanctioned by tradition. Think of the Christian church. For centuries churches were built to signify the City in Heaven of which Scripture speaks, an ideal architecture thought to do full justice to the requirements of human dwelling. This does not mean that signification could mean more or less literal representation. How could necessarily material buildings, belonging to the earth and to time, imitate an ideal, spiritual and timeless architecture? To establish his buildings as signs of this ideal, the architect had to begin with the architecture he was familiar with, with inherited ideas of what a house, a city, a church should look like. To build a church that would not only function as a church, but would represent a church by signifying the Heavenly City, the builder had to represent terrestrial paradigms. The intended meaning governed the mode of representation. Truth to the spiritual meaning of the church demanded that spiritualization of matter effected by the diaphanous walls of Gothic cathedrals and the apparent weightlessness of their vaults.

But what matters is the general pattern. By representing and transforming other architecture, the architect's work gestures towards an ideal architecture. The nature of this ideal governs the way in which the architect takes up and transforms what others have built before him. In this sense architecture may be said to be doubly representational. First of all it represents other architecture, perhaps some paradigmatic structure or structures, more often some inherited schema, exemplified by many structures. What is represented and the way a building takes up, transforms, and denies its precursors, lets it represent an ideal building—and gesture towards an ideal dwelling. Truth in architecture may be understood as adequacy to this ideal.

Take the Bauhaus's response to the past. Its rhetoric of truth to function and materials speaks of an ideal of dwelling unburdened by the ideological baggage of history, obedient only to reason and to the demands of human nature. By now we have learned that we cannot step outside history and that "human nature" is not a given but a precarious compromise between what nature and history have fixed and what remains to be established. We have become suspicious of those who in the name of truth would prescribe how we are to build and dwell.

11. Plocek House, Warren, N.J., Michael Graves, 1977–80. Main Entrance.

12. "Interesting" Portland Public Services Building, Michael Graves, 1980–82.

13. "False" Beneficial Center, New Jersey, Hillier Group, 1983.

But what happens when ideals are lacking or too empty or vague to provide our appropriation of the past with a direction? To counter this lack of direction, the architect may turn to the lie, seek truth and salvation in timid observation of "the grammatical laws" of some past style. Such attempts are necessarily in bad faith. Historicist architecture escapes the architectural lie only to the extent that its appropriation of the past asserts a living ideal.

More in keeping with the subjectivism of modernity is another strategy. Why not consider the architecture of the past a reservoir of material for aesthetic play? Much postmodern architecture engages in such play. Often such play will involve the dislocation and re-presentation of elements of some past vocabulary. To demand truth of such play makes little sense. What matters is rather that it be interesting and what makes an architect's work interesting is not obedience to an ideal or some inherited convention, but the ability to stay just one step ahead of what is currently accepted and expected. It is in the nature of the interesting that it will seek out journalism. To stay ahead of what is currently fashionable, the architect needs to keep in touch with the institutions that report architectural fashions. A history of postmodern architecture should be written from the point of view of the interesting.

For truth in architecture the interesting substitutes witty departures from what has come to be expected. One difficulty with such work is that it loses its appeal as expectations change. The interesting is bound to the present. It entertains and distracts, but does not allow for a genuine dwelling.

Another difficulty is that such departures are likely to seem arbitrary. To escape arbitrariness the work must so use what it takes from the past that the resultant work appears to justify itself. Using, while yet resisting, a view of architecture as an art of representation, we may therefore want to insist that works of architecture represent only themselves. In this view, truth in architecture becomes truth of self-representation, becomes aesthetic coherence. We have returned to the aesthetic understanding of architecture mentioned in the first section.

VI

I began with Ruskin's lament that architecture had succumbed to the lie. But if works of architecture can lie or tell the truth, they must be in some way like assertions. I suggested that what they assert is their own being as architecture. Following Ruskin's lead, we may want to call such assertions false when there is no agreement between a building and what we take to be the essence of architecture, either because that essence has been misunderstood or because, although understood, the building yet fails to meet its demands; lies when such misunderstanding is willful or agreement only pretended.

Such talk, however, presupposes that there is such a thing as the essence of architecture and that it has been, or at least can be, correctly determined. But is this essence not a fiction? Have not all proposed determinations of the essence of architecture been the products of cultural prejudice, which, when it appeals to "truth" and "essence," claims an authority that does not belong to it? Just as ideal images of human existence have changed, so have ideals of building and the readings architects have offered of what matters in architecture. The breakdown of the theocentric architectural system of the Middle Ages and its replacement by the anthropocentric system of the Renaissance illustrates such change; so does the breakdown of the latter in the eighteenth century. Since that time rationalism, aestheticism, and historicism have labored in vain to fashion a similarly comprehensive synthesis. History invites us to relativize "truth" and to replace "essence" with "ruling convention."

This need not be regretted; human being is essentially underway, in search of its own being. Just because "truth" suggests "timeless validity," appeals to truth in architecture are likely to violate the openness that is inseparable from the human situation. All too often such appeals are born, not of error, but of a fear of uncertainty that would gladly sacrifice creativity to repetition, the future to the past. Invoking "truth," such fear perverts truth into lie.

But as my last use of "lie" argues, we do not have to give up all talk of truth or of the essence of architecture. To be sure, the essence of architecture is an elusive form that receives different contents as cultural conditions change. But that form is not, therefore, a mere fiction. No matter what contents it receives, architecture will always have to provide both physical and psychological shelter. Physical shelter demands that architecture heed the Vitruvian commandments of durability and convenience; psychological, that it speak to human beings of their place, or rather, of their way. To speak, architecture must present itself as architecture, must represent architecture. Beauty in architecture is inseparable from such representation. Architecture that represents architecture inevitably also interprets what architecture has been and of what it is to be. Interpretation of the latter guides the former, interpretation of the former communicates the latter.

Such interpretation will carry conviction only if the communication is effective and allows those to whom it is addressed to recognize in the architect's work an ideal that answers to their needs. Ideals are human constructions, creative readings of the human situation. Architecture, too, contributes such readings. But creativity does not mean arbitrariness. Architecture is constrained by the demands placed on it by human nature, which may not be understood as an abstract, timeless given, but as concrete and evolving in time. True architecture answers to, shapes, and helps us recognize these demands. Architecture, too, edifies. The term *edifies* has not enjoyed a good press in recent years. That is unfortunate. It invites reflections that could recall architecture to its truth.

NOTES

1. John Ruskin, *The Seven Lamps of Architecture* (New York: Noonday Press, 1974), p. 35.

2. Ruskin, *Seven Lamps*, pp. 68–69.

3. Robert Venturi, Denise Scott Brown, Steven Izenour, *Learning from Las Vegas: The Forgotten Symbolism of Architectural Form* (Cambridge, Mass.: The MIT Press, 1977), p. 87.

4. Ruskin, *Seven Lamps*, p. 197.

5. William Hubbard, *Complicity and Conviction: Steps toward an Architecture of Convention* (Cambridge, Mass.: The MIT Press, 1981).

6. Ruskin, *Seven Lamps*, p. 196.

7. Friedrich Nietzsche, "On Truth and Lie in an Extra-Moral Sense," *The Portable Nietzsche*, tr. Walter Kaufmann (New York: Viking Press, 1954), p. 47.

8. Emil Kaufmann, *Architecture in the Age of Reason* (New York: Dover, 1968), p. 96.

9. Ruskin, *Seven Lamps*, p. 115.

10. Emil Kaufmann, *Architecture in the Age of Reason*, p. 97.

11. Ruskin, *Seven Lamps*, p. 38.

12. Arthur Schopenhauer, *The World as Will and Representation*, Vol. I, tr. E. F. J. Payne (New York: Dover, 1969), p. 215.

13. Ruskin, *Seven Lamps*, p. 40.

14. Martin Heidegger, "The Origin of the Work of Art," *Poetry, Language, Thought*, tr. Albert Hofstadter (New York: Harper and Row, 1971), p. 46.

15. Heidegger, "The Origin of the Work of Art," p. 46.

16. Jacques Maritain, *Art and Scholasticism*, tr. Joseph W. Evans (New York: Scribner's, 1962), p. 52.

17. Friedrich Nietzsche, *The Will to Power*, tr. Walter Kaufmann (New York: Vintage Books, 1968), par. 822, p. 435.

18. Ruskin, *Seven Lamps*, p. 41.

19. Ruskin, p. 44.

20. Ruskin, p. 44.

21. Hermann Bauer, *Rocaille; Zur Herkunft und zum Wesen eines Ornament-Motifs* (Berlin: De Gruyter, 1962), pp. 71–72.

1. Entrance to the masjid at Abiquiu.

HASSAN FATHY IN NEW MEXICO

Abdullah Nuridin Durkee

Egyptian architect Hassan Fathy is known to us primarily through his book Architecture for the Poor. *In this work he recounts his radical decision in the nineteen forties to forgo modern technology and architecture in favor of ancient Nubian mud brick construction and the traditional forms of his people. Beginning with the design and construction of the village of New Gourna, he has spent his life demonstrating both the beauty and the utility of this approach. While Fathy has been honored more in word than in deed, he has become the mentor of a small but significant group of architects who strive to build with an architecture and a technology that is appropriate to the people who must use it. For his lifetime work, Fathy was honored with a special chairman's award by the Aga Khan award for architecture, in 1980. The following article is an account of Fathy's first and only project to be built in North America, as told by the builder Abdullah Nuridin Durkee. The article is accompanied by the words of Hassan Fathy edited (except as otherwise noted) from a series of talks he gave in New Mexico in 1980.*

CURRENTLY UNDER CONSTRUCTION in the Chama River Valley, near Abiquiu, New Mexico, is a project of the Dar al-Islam Foundation. It was conceived not only as an agrarian and craft-making village capable of supporting a small (100 families) community of Muslims but also as a religious, educational, and cultural center that could be a seed for future projects in North America. The architecture of the buildings, as designed by the eminent Egyptian architect Hassan Fathy, reflects traditional Islamic values in plan, form, and detail. Under the instruction of Fathy and two skilled Nubian master builders, the buildings are constructed according to ancient building technologies, using the humblest of materials: mud, straw, and water. While its roots are ancient, Dar al-Islam is at the forefront of an evolving technology that is responsive to human values.

The relative ignorance of Islam by American society as a whole and the misunderstandings that can result from it have been amply demonstrated in recent years. But for American Muslims perhaps an even greater problem has been the lack of traditional Islamic educational facilities that could reinforce their own understanding of Islam. Isolated socially, culturally, and ethnically from each other and from their roots, they have little exposure to the breadth of Islam as a tradition and a way of life. Dar al-Islam was founded in an attempt to alleviate both of these problems.

In a sense Dar al-Islam arose out of a fortuitous encounter with a Saudi engineer and industrialist, Mr. Sahl Kabbani, while the author and his wife were studying in Makka (Mecca) in 1979. With Mr. Kabbani's advice, as well as with the financial assistance of Princess Moothie bint Khalid and Princess Johara bint Khalid, the Dar al-Islam Foundation was formed in November 1979. After an extensive search, a suitable site was located in the Chama River Valley of New Mexico, fifty miles north of Santa Fe, near the old American Indian and later Spanish-American village of Abiquiu.

In addition to its physical assets, that is, the necessary land, water, and soil for cultivation, construction, and habitation, the Abiquiu site has geographic, climatic, and cultural characteristics that make it ideal for an Islamic center. Located at the same latitude as

2. Taos Pueblo, Taos, New Mexico. Typical of traditional adobe architecture of the Southwest, rooms and openings are spanned with wood beams.

3. San Xavier del Bac, near Tucson, Arizona (1783–1797). This was one of the most elaborate Spanish baroque buildings in the Southwest, and the last constructed with adobe vaults and domes.

many of the major Islamic regions of the world, it also shares a similar topography, climate, and a long history of mud brick architecture (Fig. 2).[1] The native Americans of northern New Mexico already had an ancient tradition of poured earth construction when the Spanish settlers introduced the adobe brick. The Spanish had in turn learned this technology from the Moors (Moroccans), who had learned it from the Arabs. Even the name adobe derives from *atube*, the Arabic word for brick. The Spanish also carried the knowledge of brick vault and dome construction to the New World, but it did not survive in the adobe architecture of the Southwest. The last building by the Spanish employing this technique was the Mission of San Xavier del Bac, in what is now Arizona, completed in 1797 (Fig. 3).

The two-square-mile property was purchased with a generous grant from the late King Khalid of Saudi Arabia, and planning of the project commenced (Fig. 4). The first priority was construction of a place of worship (a *masjid*), to be followed by a school (*madressa*) for the Muslim children and quarters for students and teachers (*riwaq*). Construction of housing and development of shops, small industries, and agriculture were to commence simultaneously. Later phases would include construction of a clinic, a women's center, an administration building, a library and an Islamic Institute for adults. Dr. Hassan Fathy readily agreed to be architect and master planner for Dar al-Islam. He was very enthusiastic about the possibilities for such a project in the United States, particularly in the Santa Fe area, which is known for its adobe architecture. Furthermore, the Southwest is in the forefront of innovative ideas in appropriate technology, to which Dr. Fathy is deeply committed.

The first phase of the project was to build a small *masjid* (mosque). This would not only provide a place to worship and a practical means of learning the techniques of dome and vault construction but would also give local people a positive introduction to the project by way of a "hands on" workshop. Dr. Fathy agreed to produce the plans for the masjid without cost and to come to Abiquiu when the preliminary work was completed to conduct a workshop in dome and vault construction (Fig. 5).

Work commenced on the masjid in June of 1980 under the supervision of local Muslim builders. There was some difficulty in getting acceptance of the plans by the state building code officers, because no one had built adobe domes or vaults in North America since the art had died out in the seventeenth century. The problem was overcome when engineer Robert Kraus of Santa Fe, upon seeing studies of load tests and stress evaluation conducted at Ain Shems University in Cairo, pronounced the construction to be safe.

In September 1980 Dr. Fathy arrived in Abiquiu with two *muallims* (master builders) from Upper Egypt, Allahadin Mustapha Mohammed and Mu-

Culture is more than what goes into construction, and this is why we cannot simply import methods and materials that may seem, scientifically, to be efficient. Culture is the unique human response of man to his environment in his attempt to answer both physical and spiritual needs. Because both needs and environments vary in different parts of the world, culture, necessarily, is variable, too.

4. Aerial photograph of the Dar al-Islam site near Abiquiu, New Mexico, showing completed masjid and excavations for madressa.

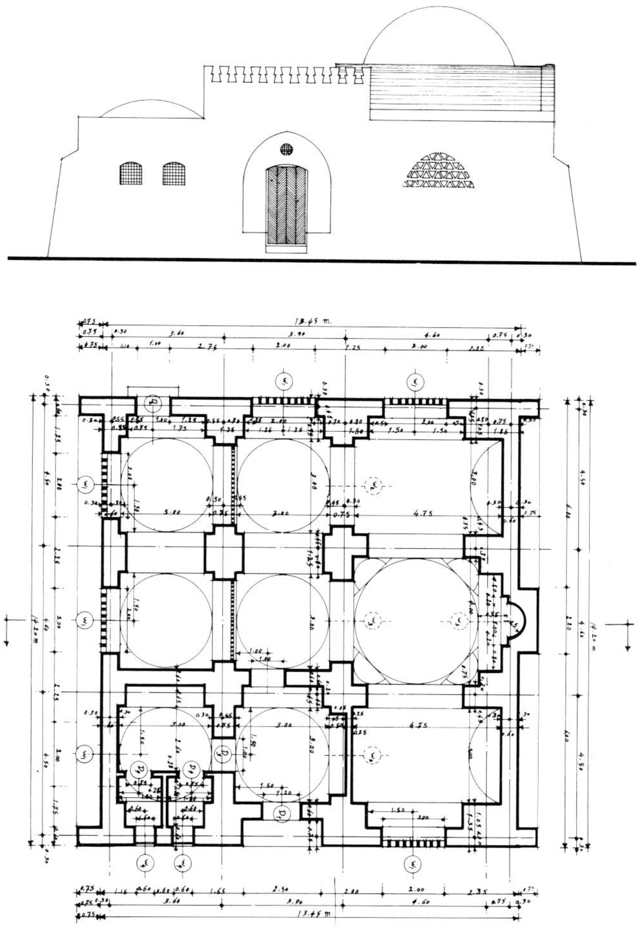

5. Plan and front elevation of masjid as designed by Hassan Fathy.

When man is handling raw material, natural material like stone, what is he doing? When he is dressing stone he is removing what is superficial and preserving what is essential. So he is spiritualizing himself and spiritualizing the stone. There is a story of a man passing three men who are dressing stone. He asked them, "What are you doing?" The first said, "I am making a living." The second said, "I am dressing stone," and the third said, "I am building a cathedral." Only the last has a sense of his task in the context of man's desires and the material's capabilities together and sees beyond merely his own purposes or the technical problem of working with the stone.

6. Hassan Fathy explains vault and dome construction at workshop at Dar al-Islam, September 1980.

7. Demonstration of Sassanid dome construction on the Abiquiu masjid.

hammad Abdul-Jalil Musa. The dome and vault workshop was attended by over two hundred people from all over the United States and abroad and received considerable publicity. But of greater importance, it trained a new corps of American builders in the traditional construction techniques (Figs. 6,7,8).

Dr. Fathy returned again in June 1981 for the dedication of the completed masjid (Fig. 9) and presented the master plan to the Board of Trustees, who accepted it in principle (Fig. 10). He also brought with him working drawings for the madressa (Fig. 11) and schematics for the riwaq. Construction of the 17,000-square-foot madressa is now nearing completion. Work is also underway on thirty units of low-cost housing designed by students and faculty of King Abd-Al Aziz University, Jeddah, and of the University of Petroleum and Minerals, Dhahran, in cooperation with Dar al-Islam. Two shops, one for carpentry, one slated for tile-making, have also been completed.

The Masjid

In order to understand the masjid, certain elements pertinent to it and to all *masajid* (plural of *masjid*) should be made clear. The function of a masjid is multidimensional. Its primary use is as a place for Muslims to gather to observe their five times daily *salat*, or prayer. It is also used as a place for learning, a place for meeting, a place for circles of remembrance, a place for reading *al-Qur'aan*, a place for travelers to rest, a place for personal retreat, and a place where, after the sunset prayer during the holy month of Ramadan, the worshipers break their fast together. To be functional masjid must accommodate all of these needs.

In Islam, prayers can be performed in any clean place on the surface of the earth. At the time of the Prophet Muhammad (PBUH)[2], the masjid was simply an enclosed space open to the sky with a covered part for prayers as a protection against the hot sun of Arabia. With the spread of Islam into areas of great antiquity, such as Egypt, Mesopotamia, and Persia, this original form was modified by contact with these older cultures and traditions. The basic form remains the same to this day, however. It is a rectangle with an open courtyard, climate permitting. The long side of the rectangle is a wall containing a *qiblah* or *mihrab* (niche) indicating the direction of Makka (Fig. 12). Included in the masjid area is a place for performing

In nature, no two men are alike. Even if they are twins and physically identical, they differ in their dreams. The architecture of the house emerges from the dream; this is why in villages built by their inhabitants, we will find no two houses identical. This variety grew naturally as men designed and built their thousands of dwellings through the millenia. But when the architect is faced with the job of designing a thousand houses at one time, rather than dream for the thousand whom he must shelter, he designs one house and puts three zeroes to its right, denying creativity to himself and humanity to man.—"Human Settlements in Newly Reclaimed Lands"

8. Partially completed masjid, with pointed arch over entrance (left) and Sassanid dome flanked by catenary vaults marking the prayer hall (right).

9. Completed masjid, showing cresting above entrance and brick claustrawork in window openings.

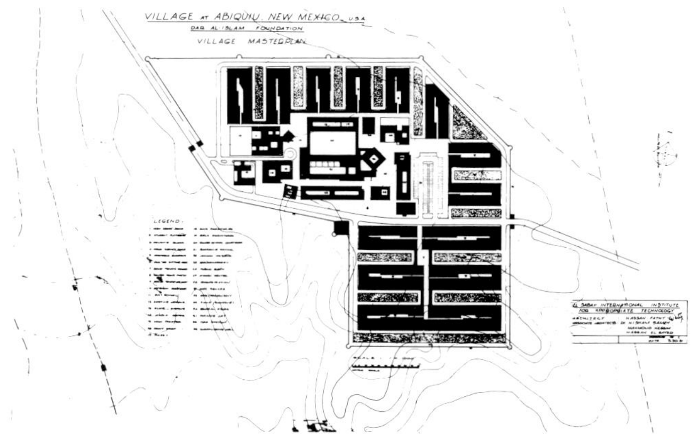

10. Hassan Fathy's master plan for the completed Lar al-Islam village. Masjid and attached madressa are at left.

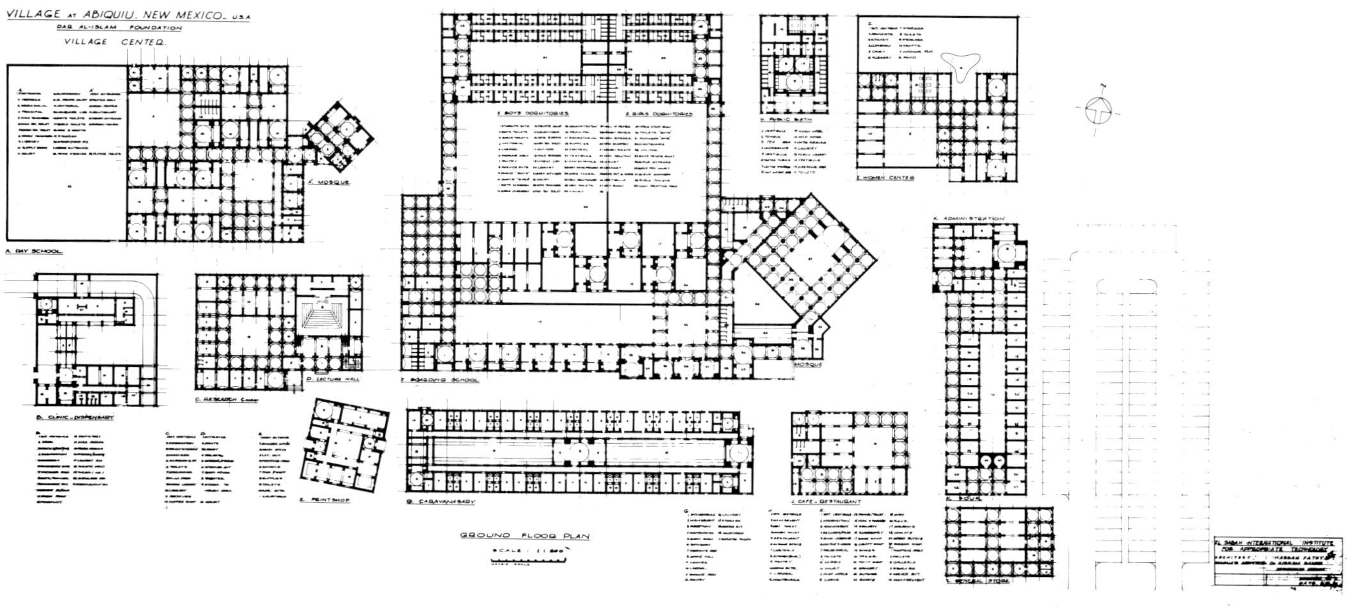

11. Plan of village center at Dar al-Islam. Madressa and completed masjid (oriented to Makka) are at left.

12. Northeast wall of masjid. Qiblah oriented toward Makka protrudes from wall below Sassanid dome of prayer hall.

13. Aerial view of completed masjid and concrete foundations of madressa.

the ablution necessary in the observance of salat.[3]

The masjid is also used as a gathering place and a place for teaching. It is in these functions that we can see the wisdom of its proportions and the human scale of its form. In pre-industrial times the size of the *iwans*, or bays, was determined by the functional spanning ability of the available timbers or by the limits of mud brick vaults and domes. They were usually built in modules of 3 m to 4.5 m, that is, sufficient to accommodate two to three rows of prostrate believers. While each city would have at least one very large masjid, the majority of masajid, located in villages or neighborhoods, or at rural crossroads, were quite intimate in scale. The very limitations of the materials and techniques used ensured respect for the human scale. The repetitive structure of pillars and arches formed iwans, cells of space ideal for classes of fifteen to twenty students sitting in circles around a teacher. The cellular form allowed many classes to be held simultaneously, each in a separate iwan, thus maintaining the integrity of each circle and the efficacy of face-to-face instruction. While classes might be in progress in several iwans, another might hold a solitary reader, while another might hold a group discussion, while in a bay to the rear a traveler might be resting prior to continuing his journey. The cellular form gave to each of these diverse activities a sense of intimacy and privacy. At the call to prayer all of these separate groups once again re-form into a united congregation assembled in rows behind the Imam. Thus, the masjid is at once a space for all, which provides all with a space.

Many masajid, including the one in Abiquiu, contain a separate space for women worshipers.[4] The space is separated by a wood or plaster screen that is pierced in such a way that the leader of prayer can be seen and heard and yet the privacy of the women is maintained. In masajid where a separate space is not provided, curtains are hung to shield the women from the eyes of men unrelated to them.

The minaret was a later addition to the masjid. In early days the *muezzin* (caller to prayer) simply stood on the roof to call the faithful to salat. This is the present situation in Abiquiu. Plans have been made to add a minaret to the masjid in the near future. The function of the minaret is to provide a high place from which the call to prayer can go out unobstructed in all directions. Its symbolic function is to express the aspiration toward the divine and to provide the complementary vertical axis to the horizontal façade of the masjid. There is a story concerning the Prophet Muhammad (PBUH) being asked by a man, "What is life and what is hope?" Muhammad (PBUH) took his stick and drew a circle on the earth, saying, "This is life." Then he drew a straight line going from the center and crossing the circumference and going beyond it, saying, "This is hope." If the dome of the masjid is the expression of the circle of life, then it will be seen that the line of the minaret against the sky is the expression of hope.

The small masjid at Abiquiu was designed to serve in its construction phase as a school for teaching mud brick building techniques and, once completed, to form part of a madressa (primary school) complex. At present it is used by all members of the community, but in time it will become necessary to build a larger masjid. The new masjid, plans for which have been drawn by Hassan Fathy, will be a further exploration of the same forms. Indeed, if one looks at the master plan one will see that all of the buildings–masajid, madressa, riwaq (boarding facility), *souk* (market), *fonduk* (hotel) and houses–are endless replications and articulations of the basic vocabulary of arch, vault, and dome.

Technical architecture must be subordinated to the natural principles of harmony and reaction to forces, as well as to the human principles of tradition, human scale, and spirituality. Technique, which can be taught in the schools, is not the same as art, which is a gift of God. Technique is like musical scales. Scales are not music by themslves. If we used technique mechanically, it would be like going to a concert and hearing someone playing the scales. There must be more–and when there is, we hear music, we see harmony, we create culture. And when technique is given to different people, just as when the scales are used by different people, the results will differ, because the cultural background of each individual or group will differ.

The Construction

The methods employed in the construction of the Abiquiu Masjid and on the madressa are almost exactly the same as those employed in the Middle East and, with the exception of the domes and vaults, the same as those still employed in northern New Mexico. The major differences involve the use of labor-saving devices. Mud is mixed in a large mixer cannibalized from an old cement truck to eliminate the backbreaking task of hand mixing, and pallets of bricks and "boats" of mud are delivered to high walls and roofs by fork lift rather than by human labor. No other mechanical aids are used. The other major concession to western technology was the pouring of concrete foundations (Fig. 13) on which to lay the mud brick walls in distinction to the stone and rubble fill foundation traditionally employed.

By praising tradition I do not suggest that we never change; that would be an impossibility. But we do find architectural elements that are as valid today as they were yesterday, because the forces that underlie the process of the creation of form are constant, and should therefore be preserved. Other elements, on the other hand, may still be valid in design and concept, but the form or size may have to be altered because changes in material or socioeconomic conditions; these elements have to be transposed. It is in this case that modern science might make a contribution to culture. But only those elements that have outlived their time and are antiquated should be eliminated.

The Bricks

The building of modular and cellular forms is dependent first of all upon the smallest modules used, the bricks. In the Abiquiu masjid three sizes of bricks were used. To supplement the standard adobe brick traditionally used in the Southwest two "new" sizes of bricks were introduced from the Middle East. Because of their smaller size these enable the bricklayer to articulate better the curves of dome and vault as well as to lessen his physical exertion. They also reduce the weight of the "green" courses of the domes and vaults.

Every material imposes its forms in architecture, according to the resistance, according to the manipulation of the material, and so on. In other words, the material itself is one of the forces determining forms, but it always works in an interaction with other forces, natural and human, that establish a given culture.

The lower courses, up to the height where the arches spring (1.40 m), are built of the standard New Mexico brick, 10 in. × 14 in. × 4 in. Above that height the size shifts. Vaults and domes are laid up with bricks 10 in. × 6 in. × 2 in. The thinness and the lightness of these bricks allow the articulation of the curves required. The other size used is 10 in. × 5 in. × 3 in. These bricks are used in arches and in the higher reaches of the building, where raising and moving the larger (30–32 pounds) bricks tends to fatigue the brick layers and slow the work. Of course the smaller

Consider the work done for a handmade structure. With the placing of every brick was the decision to put it here as an arch, there as a wall, and so on, and in the effect of every decision is the basic nature of man. There is an aesthetic value in this method of working by hand. If you take the opposite method of prefabrication and you have just these bulks that you carry with a crane and set with a crane, you see that man contributes nothing. Too often now we build in this way. We build one house and put six zeroes next to it, making it a million. But the mechanical takes away from human harmony with nature, materials, and tradition; the mechanical, like symmetry, is death.

14. Basic tools and bricks used in adobe construction. Pole and stick used for laying out radius of Byzantine dome are shown at top: Three brick sizes used are, from left to right: the flat Egyptian style brick used for vaults and domes; a larger Egyptian brick used for walls; and the standard Southwestern adobe used for walls. At bottom are tools made of metal lath on wood, used for shaping bricks and arches.

15. Mixing mud for mortar during masjid construction.

16. Madressa under construction, with walls completed up to spring point of arches. Masjid shown beyond.

bricks also require more handling to complete the same cubic footage. The two smaller bricks have been used since the time of Sumeria, when brick making was first developed (as distinct from poured mud walls), and they are still in use throughout the Middle East (Fig. 14).

As in the Middle East, bricks are made of earth combined with sand, straw, and water, and molded to the desired form. The addition of straw and cow dung to the mixture is sufficient to make both the bricks and the mortar durable and relatively resistant to rain (Fig. 15). The lower courses of the walls at Abiquiu are laid in essentially the same manner as in traditional adobe buildings of the Southwest. Header courses are alternated with running bond courses to insure cross-bonding. But at a height of 1.4 meters (4.5 feet), the springing point of the arches, the construction changes (Fig. 16).

The Pointed (3 : 5) Arch

The pointed arch is used extensively in Islamic architecture. It is constructed of two arcs whose centers lie on the springing line and which intersect at a point at the top of the arch. The rise of the arch ranges from three-fifths to three-quarters of its span. At Abiquiu what is known as the 3:5 arch is employed. The arches are used to frame doors and windows, to connect between spaces and to support the domes. The largest such arches, supporting the Sassanid domes, span three meters. In Egypt the arches would be constructed on "soft forms" of mud bricks covered with mud and sand, but at Abiquiu re-usable plywood forms have been employed (Fig. 17,18).

The areas of wall between arches are infilled with standard adobe brick coursing to just above the point of the arches. Above this level three main forms are used to enclose spaces. These are the catenary vault, the Byzantine dome and the Sassanid dome (Fig. 19).

The Catenary Vault

The method of building catenary vaults of mud brick was re-discovered by Hassan Fathy in 1941. During a trip to Nubia (Fig. 20), in Upper Egypt, he saw the Fatimid Cemetery in Aswan and the tenth century Coptic Monastery of St. Simeon, both of which employed mud brick vaults. At Luxor he saw the nineteenth dynasty granaries of Ramesseum, long mud-vaulted storehouses that were still standing after 3,400 years. The problem was to find craftsmen who still had the knowledge to build them. In Nubia Hassan Fathy found two superb craftsmen, Abdul-Hamid and Abu-Noor, who still practiced the art of vault-making. An intense period of building began with Abdul-Hamid and Abu-Noor laying up vaults in all kinds of experimental situations. The vaults stood, they were cheap, and the work went quickly. A 3 m × 4 m room with a span of 3 m could be finished in a

17. *Plywood forms used in construction of 3 : 5 pointed arches.*

18. *Completed arches at madressa.*

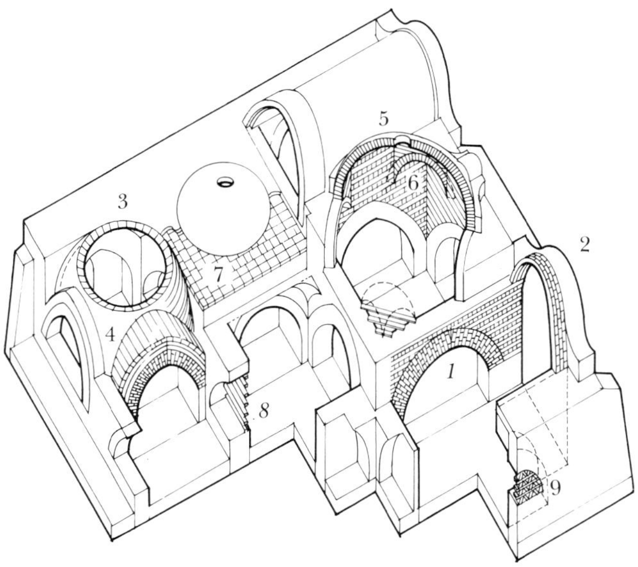

19. *Cutaway drawing of Abiquiu masjid showing the basic types of construction: the 3 : 5 pointed arch (1); the catenary vault (2); the Byzantine dome (3) constructed on pendentives (4); and the Sassanid dome (5) constructed on squinches (6). Also shown are roof construction (7), wood lattice (8) between interior spaces, and brick claustrawork in openings to the outside (9).*

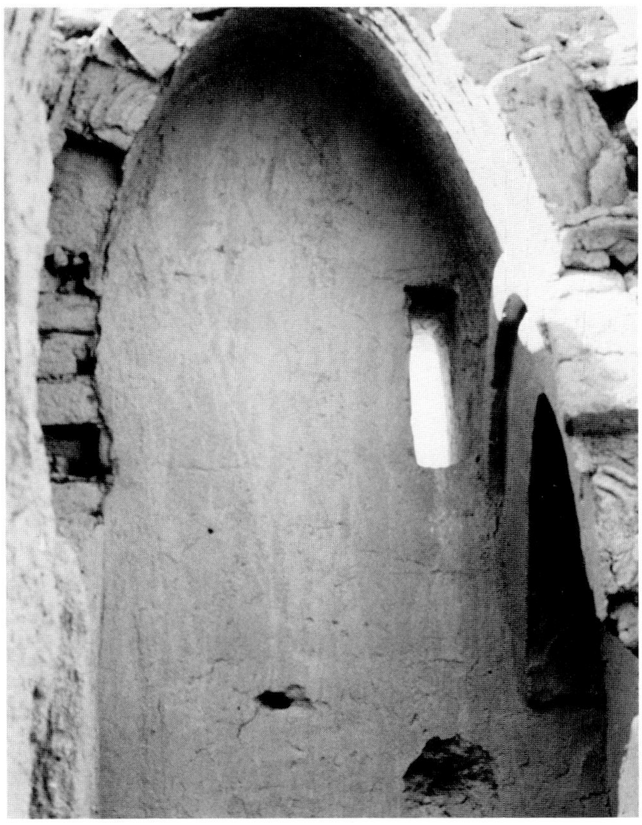

20. *Bagawat cemetery, Kharga Oasis, second century A.D.*

21. Nubian mason constructing catenary vault at Giza, Egypt in 1983.

22. House of Dr. Hamid Said, designed and constructed by Hassan Fathy in 1942.

23. Bariz village center, showing partially constructed inclined catenary vaults. Designed by Hassan Fathy, constructed 1967.

24. Construction of a Byzantine dome with pendentives infilling between arches.

day and a half. A twelfth dynasty Egyptian technique had re-emerged, and an architectural renewal began (Figs. 21–23).

The vaults at Abiquiu are built according to this method, as described in Hassan Fathy's book *Architecture for the Poor.*[5] This description can give an idea of how the vaults are constructed but, as demonstrated at Abiquiu, the technique can only be learned "hands on." It must be transmitted by a living, breathing, bricklaying master, as in truth all of these techniques must be learned by the hands and the eye and under the hands and the eye of one who has mastered them.

The Byzantine Dome

The Byzantine dome actually should be called the "Ancient Egyptian Dome," since there is still one standing in the tomb of Seneb in the cemetery at Giza. It is basically a hemisphere that rises on pendentives set into the interstices of four arches set in a square at 90 degrees to one another. The method for determining the radius and eventual height is to set a post, the top of which is level with the bottom of the interstice where the arches join, and fix to this post a stick whose length is determined by the distance of the center to the interstices. This stick is attached to the post by a hinged pivot. This provides a guide by which the bricks are then laid up in concentric circles to form a dome (Figs. 24,25).

The Sassanid Dome

The Sassanid dome, as its name implies, derives from Persia. In this configuration the corners where the arches meet are filled square until leveled off at the top of the arches. This forms a square. Squinches are then set at all four corners, turning the square into an octagon. The post is then set up so that its top is level with the top of the square room and the radius stick is attached as before. The bricks are then laid up until the height of the resulting octagon reaches the height of the squinches. At this point, depending on the size the dome (3 m or 4 m to 4.5 m), six or eight more smaller squinches are added, creating a fairly round shape, i.e., a polygon of twelve or sixteen sides. The circle of the hemisphere then rises harmoniously from this point (Figs. 26,27,28).

25. Construction of a pendentive.

26. Construction of masjid. Pointed arches in foreground form base for Byzantine dome. Squinches under construction in background form base for Sassanid dome.

27. Interior of completed Sassanid dome before plastering.

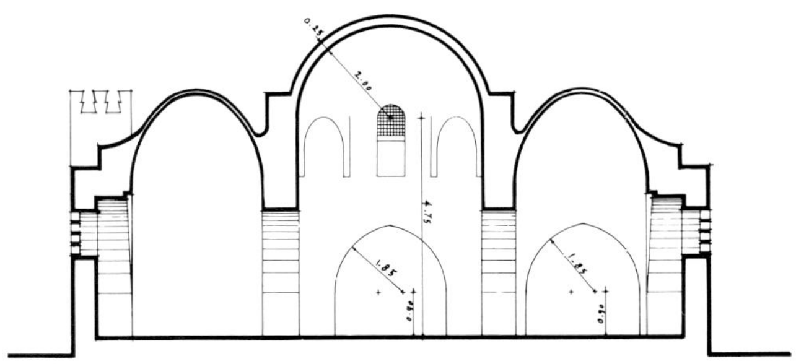

28. Section of masjid, through qiblah, Sassanid dome and catenary vaults.

Building is a creative activity in which the decisive moment is the instant of conception, that instant when the spirit takes shape and all the features of the new creation are virtually determined. While the characteristics of a living creature are irrevocably settled at the moment of fertilization, the characteristics of a building are determined by the whole complex of decisions made by everyone that has a say in the matter, at every stage of its construction. Thus the instant of conception on which the final form of a living creature rests becomes for a building a multiplicity of such instants, each playing a decisive part in the total creative process.—Architecture for the Poor

Man cannot create style overnight; this requires the communication of generations, one after another. In this manner a people creates its own forms and its own architecture, and it has its own truth, which is true to the environment. This is what each society needs. International architecture? This is international disrespect of art and architecture, and international loss of culture.

29. Sassanid dome and catenary vaults with mud plaster finish. Scupper carries water off roof.

30. Inside prayer hall of masjid.

Exterior Finishes

Once the domes and vaults are completed the spaces between them, particularly around the pendentives and squinches, are filled with broken adobe bricks. These are topped with a flat layer of whole bricks and a smooth coat of mud. All roof surfaces are sloped slightly to drain water toward perimeter roof scuppers (Fig. 29).

The traditional exterior finish of the walls, as well as vaults and domes, is a mud plaster. This, however, is labor and maintenance intensive. Accordingly, in New Mexico a surfacing material of crushed limestone aggregate, wood pulp fibers (for elasticity), and pulverized granite (for coloring) in a lactic acid base has been employed. This provides a hard but elastic coating on the building and protects the mud surface underneath.

Interior Finishes

Interior surfaces are finished with a gypsum-based plaster. It is either floated, yielding a grainy, pebbly surface, or hard-troweled to yield a smooth finish. The finish is varied with the intent of producing a light-absorbing surface (float) or light-reflecting surface (hard trowel). The surfaces are then painted, usually in an off-white pale sandy color that matches the color of the prevailing wind-blown dust that tends to catch on them (Fig. 30).

Originally the plaster was applied directly onto the adobe surface, which has lots of "tooth," but it has proven better to cover the surface of the adobes with metal lath prior to plastering. Due to the tremendous mass of an adobe structure there is always settling, especially in the first two or three years. The metal lath seems to prevent the large latitudinal cracks or buckles which result from this settling process. Ideally, adobe structures should also be allowed to "breathe" for a year before being plastered to allow moisture to wick out. The preferred interior plaster finish recipe is derived from one used over the last five thousand years in the Valley of the Kings in Upper Egypt:

> Mix gypsum with water until it loses its setting power. Add one part of gum arabic to ten of gypsum until it reaches the consistency of thin yogurt or buttermilk. Apply with brush over a two coat mud plaster base, adding powdered tempera for coloring.

Tile is used as a floor covering wherever people are likely to be wearing shoes. The tiles, most com-

31. Ablution area of masjid, showing tile and brick claustrawork.

In the ideal architecture, modeled by the craftsmen, by the artisans, man would be putting something from himself into the material and radiating something, giving it something that would be re-radiated to man, along with the essence of the material (and doing him good). The principle of accretion allows for the constant interaction of man and material and environment. To my mind architecture is like the shell of the snail, the soft part secreting calcium carbonates and by natural forces making the form by movement and surface tension. The grains would arrange themselves around this village, returning on the soft, living part that created it and gave it form. For man, the universe, the sky, the sun, the moon, the wind, his own body, his brain, his heart, and all his reactions are transferred to his fingers, shaping the brick, shaping the mud, sculpting; he is radiating from his body into the material as the environment has radiated into him. When he creates a form, it becomes a part of his environment, and will re-radiate the feelings he invested in it, which in their turn are derived from the universe around him; so man works in harmony with his world.

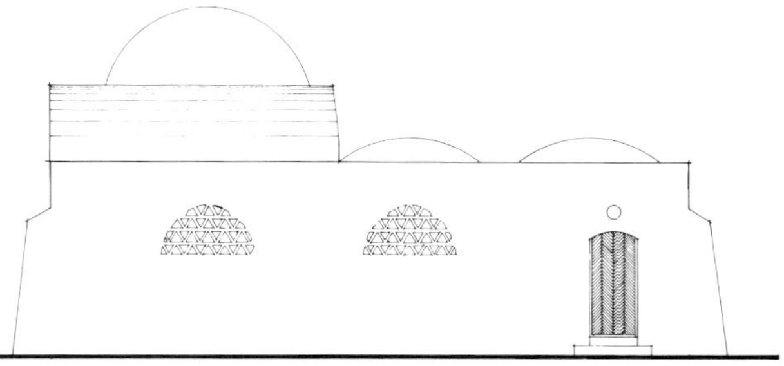

32. Elevation of masjid, showing claustrawork in window openings.

monly octagonal low-fired tiles from Mexico, are laid in a permabond "soup" on a concrete slab. The interstices are infilled with low-fired earth glaze "Talavera" tile, usually in an eight point star pattern.

Wall surfaces that come into contact with water, as in the ablution areas, are tiled with a mix of high-fired commercial tile and the same low-fired "Talavera" tile. All patterns are edged with complementary color strip borders and then bull-nosed back into the plaster (Fig. 31). In the school the same technique will be used to protect the walls from the dirty hands and trailing crayons of the young children. Tiles are also used around the heat sources to store and reradiate heat following the principle used in the Austrian kachelofen. There are now plans to build a shop for the production of tiles in a wider variety of shapes than are commercially available in the United States.

Claustrawork

Brick claustrawork is used primarily in windows, though sometimes it is used to infill arches which separate spaces of different functions. The most commonly used pattern is known as *saba-thamaniya* (Arabic: sevens and eights). The interstices of these lattices are infilled with various clear, translucent or colored glasses, creating patterns of light and shadow and color and transparency. The depth of the brick controls the amount of direct sun that enters the interior space. The shallower the brick the more direct sunlight may enter, while the deeper the brick the more light catches and bounces on the depth of the edges. This manipulation of light creates beautiful patterns both seen through the windows and cast by them on walls and floors (Fig. 32).

Woodwork

The section of open arches between the men's and women's sections of the masjid is to be filled with a rounded wood latticework known as *mushrabiya* (as opposed to the flat latticework known as *shareikha*). There are literally hundreds of geometric patterns that are traditionally used. This latticework provides for the privacy of the people on both sides while it allows free circulation of air and sound (Fig. 33).

The wood doors are similarly constructed in the *shish* or *shareikha* form. These can be composed either in purely geometric patterns or in the form of Arabic known as Kufic, which lends itself to geometric form. Thus the main doors of the masjid, which appear to be geometric, actually say, "In the Name of Allah, the

33. *Traditional mushrabiya by carpenter Hassan el Niger. House designed by Hassan Fathy, in Dashure, Egypt, 1979.*

34. *Masjid entrance showing door construction.*

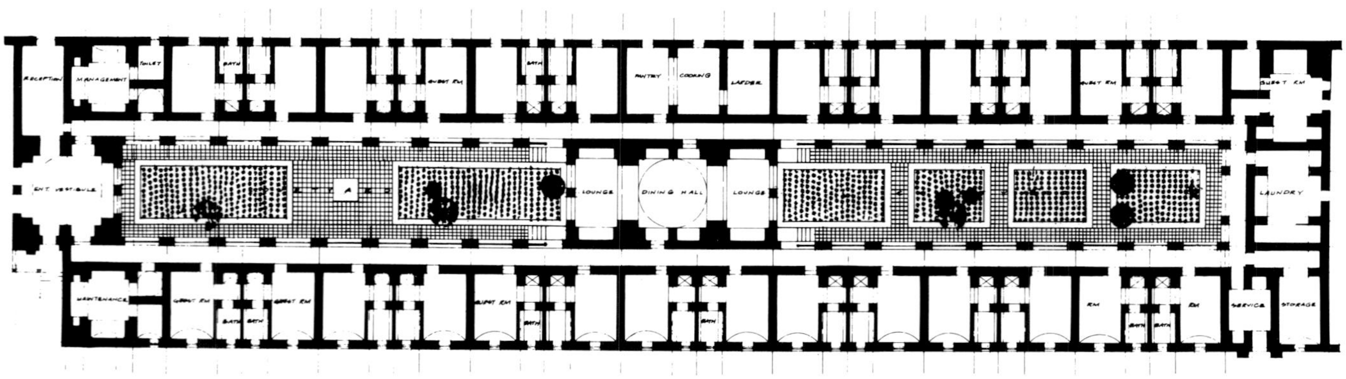

35. *Plan of caravansary.*

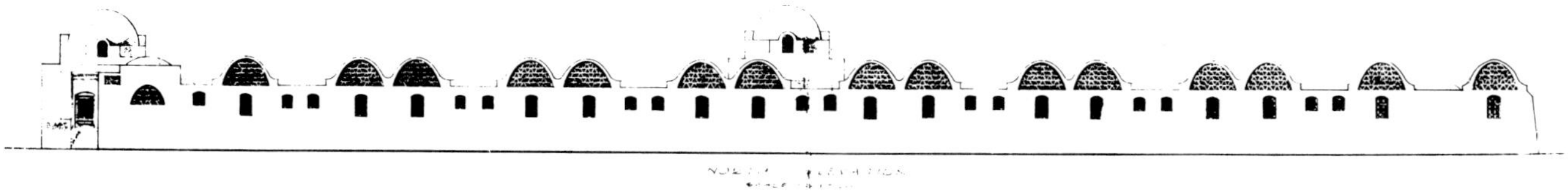

36. Elevations of caravansary.

Merciful, the Compassionate." Another form employed in the masjid is the "sabra" door. This is built up of chevron staves morticed into the stiles and rails, thus permitting the majority of the door to be built of short lengths of wood. This is a boon in desert areas where wood is in short supply or trees are so small that it is hard to produce long lengths of wood (Fig. 34).

The Canons of Building

All of these elements can form and re-form endless permutations (Figs. 35,36). The knowledge of the canons and proportions of building is a "revealed" knowledge which is transmitted from the one who knows and has mastered the knowledge. This person was traditionally known as the Sheikh al-Banna, the master of the builders. The sheikh initiated the master craftsmen into the knowledge of the laws of harmony and proportion and the master craftsmen in turn initiated their apprentices into the knowledge of the craft and the handling of the materials. The result of this cooperation manifested and manifests itself in a guild system which in turn was often a Sufi tariq. This assured that the spiritual dimensions were maintained so that the knowledge, which was of a sacred character, did not degenerate into merely secular techniques.

In the present age this system has been lost almost everywhere. No longer is building an act of conscious cooperation between the Sheikh, Master Craftsmen and Apprentices. There now exists a new and blatantly secular order of relationships, termed Architect-Contractor-Laborer. As Hassan Fathy puts it, "The revealed knowledge of the Sage is now replaced by a machine. The result is that we see so many mosques, in almost all Muslim countries, in which the arbitrary outweighs the implicit and sacredness is lacking almost entirely . . . revealed knowledge is not given to everybody or anybody, it cannot be acquired by academic learning alone."[6]

Given the apparent ease with which Hassan Fathy manipulates form and space, employing the vocabulary of arch, vault, and dome, many have imagined that one can easily build using the same vocabulary. Certainly one can easily explain how pendentives rise and fall out of arches, and how arches shift axes and link, generating domes that rhythmically join domes to merge in vaulted corridors in which latitudinal spaces intersect longitudinal vectors, frozen music put into play by the movement of the eye of the beholder. One can show how this form of building is not about walls but about the space between walls, and how this space is an assemblage of modular and rhythmic form, internal and external, converging and diverging, now multidirectional and now unidirectional. One can say it is only a matter of 3 m × 3 m or 4 m × 4 m iwans surmounted by Byzantine or Sassanid domes with squinches that turn four sides to eight and eight to twelve or sixteen, and turn square into cube, cube into circle, and circle into sphere. But in reality this will yield no fruit.

In order to learn these methods and the secrets of these forms it is necessary to put oneself under

Tradition, as I have suggested, is a key element of culture. When the craftsman was responsible for much of the work of building, traditional art came out of the subconscious of the community. There were no architects, only master builders. The master, the owner of the house, knew what he wanted because in the tradition there were examples in the rest of the city, and the master mason or master carpenter knew what he could give. On the other hand, modern architecture comes out of the conscious mind of the individual architect. It is terrible what the modern architect has saddled himself with as a responsibility. He wants to take over the responsibility that was given in tradition to the hundreds of centuries of generations, one adding to the other. The village built by accretion is more harmonious than the one built on the drawing board; it is held together by an accumulated culture, rather than by one individual's idea of harmony.

the hand of a sheikh or master. Sadly, there are few known master builders in our time, and perhaps only one working in the exalted material of earth: Hassan Fathy, at 82 years still working out of the upper chambers of a sixteenth-century Mamluk townhouse in Cairo. Among the few remaining master craftsmen are Allahadin Mustapha Mohammed and Muhammad Abdul-Jalil Musa of Nubia, who taught the craft to apprentices at Dar al-Islam. Now the craft is being taught to others. In a sense the masjid and madressa in Abiquiu exist not only for their own functional purposes, but, in the process of their construction, as a school of traditional Islamic crafts.

Students, seekers and would-be architects should note well the following statement of Hassan Fathy: "Remember, this wisdom does not belong to a unique time, it belongs to all times. It is present today as it was yesterday and can be realized by anyone who sincerely desires it and deserves it."[7]

NOTES

1. Abiquiu, which is at 34.12 degrees north, is on roughly the same parallel as Northern Maghrib (35 degrees to 37 degrees north), Andalusia (39 degrees north), Syria (35 degrees to 36 degrees north), Iran, Afghanistan, Bokhara, Samarkand, and Sinkiang Province, all places where Islam has been eminently at home.

2. Peace be upon Him.

3. In order to enable the reader to understand the masjid in Abiquiu, as well as by extension the general form of all masajid, let us follow the believer and enter.

The Door

The door marks the *barzaq* (boundary) between the outer and the inner. In traditional masajid and the Abiquiu masjid, the entry is recessed. One "enters into" sacred space from secular space. The sacred space does not protrude outward into the secular space. The door is usually set into an arch. This is not a semicircular arch (Roman or Osirian) that rises from the earth, completes its arc, and returns to the earth, but is a pointed arch (Fig. 1) whose function is to raise the eyes upward following the two tangential lines of force that meet at the keystone. This arch in turn merges with the façade, which is surmounted by crestings that connect the building with the sky. These crestings are alternate negative and positive forms of steps that allow heaven and earth to have a meeting place.

The Entrance Hall

One enters through the door into an intermediary space where one finds an area for removing and storing shoes and an entry to a small chamber for making the ablutions necessary to purify oneself in order to perform the salat (worship). The area for ablution is tiled and has tiled seats where the believer sits in order to perform the ablutions. The measurements of the facilties (Fig. 29) for ablution, seat, water taps, splash moldings, and drains must all be designed to relate to the human body and the movements of the limbs in the act of performing ablution. When one has washed off "the dust of the world," one is ready to enter the prayer hall and perform the salat.

A low barrier at the end of the entrance hall is the final separation between the inner and the outer space. In the prayer hall beyond this barrier the floor changes from a tile surface to adobe covered by reed mats, which are in turn covered by long carpets 110 cm in width, known as *saffs*, upon which the believer prostrates during salat. No shoes are permitted beyond this point, since what lies beyond is a ritually "pure" space and one does not bring the dirt of the street into this space. As one looks into the space of the prayer hall one notices the absence of all furniture and one senses a spirit of visual silence and equilibrium. Stepping over the final barrier one enters the prayer halls of the masjid.

The Prayer Hall

One enters a large rectangular space formed by cells or bays of arches surmounted by domes. The orientation of this space is shifted 90 degrees from the axis of the entrance hall so that the longest wall may face toward Makka. The space is rectangular in shape so that the worshipers may arrange themselves in long lateral rows. Since there is a greater merit in being in the first row, much attention is given to extending the longitudinal axis to accommodate the longest possible single row. These long lines also serve to define visually the horizontal dimension of the space.

The vertical dimension is achieved through the use of massive pillars that yield to pendentives, which in turn flow into domes. These domes serve the same function as the dome of the sky did in the earliest masajid, which were partly open. Often the prayer hall would be covered by flat roofs carried by beams supported on pillars with the bay directly over the qiblah/mihrab being a raised flat roof or, as in the case of the early masjid of Ibn Toloun (Cairo 157 H./876 A.D.), surmounted by a dome. The Abiquiu masjid is similar in form (though not in scale) to the Barquq masjid, in which all the prayer iwans (bays) are covered by Byzantine domes on pendentives except for the bay in front of the qiblah/mihrab, which is covered by a Sassanid dome atop squinches.

The Qiblah or Mihrab

The qiblah (Arabic: "to accept or receive in kindness") or mihrab (place of refuge) is a niche-shaped indentation in the longitudinal wall facing Makka. Its primary purpose is to direct the worshiper correctly towards Makka. Most masajid have these two very different axes. The one relates to the site and the other is determined by the direction that the believer must face in salat toward Makka. Contrary to what many non-Muslims think, Muslims do not necessarily face east to offer their salat. A vast number of Muslims face west (those from India, Pakistan, Indonesia, etc.) and to practically every other direction of the compass. The direction of the qiblah, and thus the orientation of the masjid, is determined by plotting the arc of the great circle on the globe from the place where the worshiper is to Makka. Thus the qiblah in Abiquiu is 34 degrees 17 minutes east of true north.

As can be seen from the Arabic meaning of the word, there are other qualities of the qiblah besides its being the source of determining direction. The form of the qiblah, as noted, is generally a concave niche set into an arch. The depth of this niche varies from masjid to masjid, but it is this depth that represents for the believer the gateway to the unseen and the receptive focus of the congregation. It is important to note that it is an "empty" space, which is very typical of Islamic architecture and the outlook of Islam in general. Rather than "filling" space and projecting into space or taking up space, what is accentuated is the inner receptive void. This void, which is also the crescent form so often found in Islamic art, symbolizes the receptive nature of the believer, the openness to being filled, the empty container that in containing nothing contains everything. Indeed the whole masjid and all traditional masajid have this sense of emptiness, of tranquility, of calm and receptivity to the moving spirit. An *ayat* (verse) from al-Qur'aan in the "Light Sura" refers to this sense in the following words: "Allah is the Light of the Heavens and the Earth; the likeness of His Light is as a niche wherein is a lamp (the lamp is in a glass), the glass as it were a glittering star kindled from a Blessed Tree, an olive that is neither of the East nor of the West whose light shines though no fire touches it. Light upon Light. Allah guides to His Light whom He chooses and Allah makes parables for people and Allah has knowledge over all things." These ayat are very often inscribed around the qiblah or mihrab, and a lamp is generally placed above them.

During congregational salat the Imam (prayer leader) stands directly facing this qiblah. The worshipers form behind him in *suffat* (rows) facing the wall in which the qiblah is centrally located. They follow his movements, signals for which are both visual and, more importantly, aural. The qiblah, with its concave space, thus performs another function, which is to act as a resounding chamber reflecting these aural signals to worshipers in the rear of the masjid or in other iwans who may not be in direct visual contact with the Imam.

The Movements of the Salat

The Salat of the Muslim engages the entire body. The word *salat* is most often translated as *prayer* in English; the meaning is perhaps more correctly rendered by the word "salutations." In effect the worshiper is not praying for something but rather offering salutations to Allah who is the creator of all the worlds. The fact that the entire body participates in this act indicates that for worship to be efficacious the entire being must be engaged. In terms of the design of the masjid this means that these movements must be accommodated and space provided for their manifestation. This is also another reason that there can be no objects such as furniture, in the masjid, since they would impede the movements of the rows upon rows of believers.

The purified body of the believer engaged in the ritual act of worship moves through five basic actions, the sequences, times, and rhythms of which are canonically defined. During these movements the words of Allah are also enunciated with various verses from al-Qur'aan read from the heart and the tongue in conjunction with the movements. Briefly, these movements may be described as: *Standing*, the action that distinguishes the human from the animal; *Bowing*, the act of homage of the lesser to the greater; *Prostration*, the total self-abnegation of the relative and ephemeral individual in the face of the absolute eternal imperative; *Sitting*, the station and condition of equilibrium, rest, and peace; *Turning*, the facing to the right and to the left following the sitting to bestow the word of peace to the angels to either side, to one's immediate companions, and by extension to the whole of creation.

4. Although the masjid is perfectly open to them, traditionally women prefer to pray in their houses on most occasions (this again is a further understanding of the inner and the outer, the woman being understood as inner, which is the sacred space of the home, while the men who perforce must be out in the world use the sacred outer space, the masjid). There are, however, occasions in the course of the year, such as the special *tarawih* prayers during the month of fasting (Ramadan) and the Eid festivals, which occur twice during the year, when women commonly use the masjid.

5. "The masons asked us to make them the special kind of bricks they used for vaults. These were made with more straw than usual, for lightness. They measured 25 cm × 15 cm × 5 cm (10 in. × 6 in. × 2 in.) and were marked with two parallel diagonal grooves, drawn with the fingers from corner to corner of the largest face. These grooves were very important, for they enabled the bricks to stick to a muddy surface by suction. So we made the bricks and dried them, and a week later went down to the site. On our way down I noticed that the masons had no tools, except for their adzes. I asked them, 'Where are your trowels?' 'We don't need trowels,' they said, 'the adze is enough.'"

At the scene of our failure the walls were still standing although our attempt at a vault had collapsed. In each room there were two side walls, three meters apart, and an end wall somewhat higher against which the vault was to be built. The masons laid a couple of planks across the side walls, close to the end wall, got up on them, took up handfuls of mud, and roughly outlined an arch by plastering the mud onto the end wall. They used no measure or instrument, but by eye alone traced a perfect parabola, with its ends upon the side walls. Then, with the adze, they trimmed the mud plaster to give it a sharper outline.

Next, one at each side, they began to lay the bricks. The first brick was stood on its end on the side wall, the grooved face flat against the mud plaster of the end wall, and hammered well into this plaster. Then the mason took some mud and against the foot of this brick made a little wedge-shaped packing, so that the next course would lean slightly towards the end wall instead of standing up straight. In order to break the line of the joints between the bricks the second course started with a half-brick, on the top end of which stood a whole brick. If the joints are in a straight line, the strength of the vault is reduced and it may collapse. The mason now put in more mud packing against this second course, so that the third course would incline even more acutely from the vertical. In this way the two masons gradually built the inclined courses out, each one rising a little higher round the outline of the arch, till the two curved lines of brick met at the top. As they built each complete course, the masons were careful to insert in the gaps between the bricks composing the course (in the extrados of the voussoirs) dry packing such as stones or broken pottery. It is most important that no mud mortar be put between the ends of the bricks in each course, for mud can shrink by up to 37 percent in volume, and such shrinkage will seriously distort the parabola, so that the vault may collapse. The ends of the bricks must touch one another dry, with no mortar. At this stage the nascent vault was six brick-thicknesses long at the bottom and only one brick-thickness long at the top, so that it appeared to be leaning at a considerable angle against the end wall. Thus it presented an inclined face to lay the succeeding courses upon, so that the bricks would have plenty of support; this inclination, even without the two grooves, stopped the brick from dropping off, as might a smooth brick on a vertical face.

Thus the whole vault could be built straight out in the air, with no support or centering, with no instrument, with no drawn plan; there were just two masons standing on a plank and a boy underneath tossing up the bricks, which the masons caught dexterously in the air, then casually placed on the mud and tapped home with their adzes. It was so unbelievably simple. They worked rapidly and unconcernedly, with never a thought that what they were doing was quite a remarkable work of engineering, for these masons were working according to the laws of statics and the science of the resistance of materials with extraordinary intuitive understanding. Earth bricks cannot take bending and sheering; so the vault is made in the shape of a parabola conforming with the shape of the bending moment diagrams, thus eliminating all bending and allowing the material to work only under compression. In this way it became possible to construct the roof with the same earth bricks as for the walls. Indeed, to span three meters in mud brick is as great a technical feat, and produces the same sense of achievement, as spanning thirty meters in concrete.

The simplicity and naturalness of the method quite entranced me. Engineers and architects concerned with cheap ways of building for the masses had devised all sorts of complicated methods for constructing vaults and domes. Their problem was to keep the components in place until the structure was completed, and their solutions had ranged from odd-shaped bricks like bits of three-dimensional jigsaw puzzles, through every variety of scaffolding, to the extreme expedient of blowing up a large balloon in the shape of the required dome and spraying concrete onto that. But my builders needed nothing but an adze and a pair of hands.

Within a few days all the houses were roofed. Rooms, corridors, loggias were all covered with vaults and domes; the masons had solved every problem that had exercised me (even to building stairs). It only remained to go out and apply their methods throughout Egypt." From Hassan Fathy, *Architecture for the Poor: An Experiment in Rural Egypt* (Chicago: The University of Chicago Press, 1973), pp. 9–11.

6. Hassan Fathy, "Mosque Architecture/Sacred Space" (unpublished manuscript.

7. Ibid.

1. Aerial view of the project taken in January of 1971. The full scale mock-up used to test concrete pouring techniques for the curved roof vaults can be seen at upper right.

THE IMMEASURABLE MADE MEASURABLE: BUILDING THE KIMBELL ART MUSEUM

A. T. Seymour III

LOUIS I. KAHN has been quoted as saying, "Architecture is the embodiment of the immeasurable." However, in this writer's opinion, the Kimbell Art Museum was measured inch by inch, dollar by dollar, and day by day by the general contractor, Thos. S. Byrne, Inc. of Fort Worth, Texas.

The unusual and striking architectural design of Fort Worth's Kimbell Art Museum was inspired by its dynamic director, Dr. Richard F. Brown, and was conceived by architect Louis I. Kahn of Philadelphia, in association with Preston M. Geren, Architect & Engineer & Associates of Fort Worth. The results of their unusual requirements presented a broad variety of challenges to the general contractor.

Byrne's initial task, which began in early 1969, involved developing a realistic construction budget for Kimbell concerning all phases of the contemplated architectural, structural, mechanical, electrical, and extensive sitework improvements that had been under study by the designers since 1967. A variety of specialist manufacturers and subcontractors were encouraged to join the owner–architect–contractor team in helping to shape the challenging and innovative Kahn design into reality, and at the same time to keep the costs within an acceptable budget to be approved by the museum's Board of Trustees.

After months of extensive cost analysis, a budget was approved and a construction schedule developed for completion in early 1972. Kimbell negotiated a "guaranteed maximum cost" contract with Thos. S. Byrne, Inc. with all savings under the maximum cost accruing to the owner. This arrangement allowed site preparation and foundation work to commence in mid-1969 with price protection for the owner while complex architectural, structural, mechanical, and electrical designs were still being finalized. As the various design components materialized, firm costs were established within the budget and material orders placed in a phased program that would meet the construction schedule developed by Byrne.

During 1969, the construction industry nationwide was experiencing a high rate of inflation, averaging one per cent cost increase per month. At the same time a labor shortage was foreseen locally because of construction of the large regional airport starting in 1970. The contractors took these factors into account and programmed their major buying and job labor employment accordingly.

As the project phased from budget to building, the contractor's challenge became even more complex due to the many never-before developed ideas of Mr. Kahn's imagination. It was the contractor's responsibility to phase the construction schedule to allow maximum time for the architect to achieve the best possible solutions to complicated design problems.

The building site was a paved parking lot. The contractor chose to divide the excavation into two phases. In mid-1969, 25,000 cubic yards of earth were excavated for the building foundations, leaving the surrounding paving in place for easy access to the building perimeter during bad weather by the large cranes and concrete trucks needed to pour the concrete roof construction. In mid-1971, an additional 30,000 cubic yards, averaging eight feet deep, was excavated for parking lots and other site improvements. This two year gap also allowed the designers time to develop their impressive site development program.

Thin concrete roof shells of cycloidal cross-section form the museum roof structure. This system consists

2. *Early view of concrete column and foundation wall construction at the ground floor level.*

3. *Reinforcing steel and post-tensioning tendons in place in the gallery floor, August 1970.*

4. *Aerial view showing pouring of central gallery floor slab, October 1970. Weight lightening polystyrene blocks can be seen in position in lower right-hand portion of central area.*

5. *Roof shell formwork components awaiting assembly.*

6. *Fabrication of formwork for the "diaphragm" end walls below the roof vaults.*

7. *View of roof construction. The flat roof slabs were poured first, serving as work platforms for the pouring of adjacent roof vaults.*

of sixteen 24′ wide × 104′ long × 5″ thick reinforced and post-tensioned concrete shells separated by 10′ wide lower flat slab units, poured in place over specially designed breakaway hinged forms on high-load adjustable steel shoring. These forms consisted of three full form setups that were built and reused an average of five times each over a ten-month period to produce the concrete finish exposed to view in the gallery areas below.

An unusual feature of the Byrne-developed roof forms was the mobility of the 19′ long curved form sections, which could be "jacked down" to roll under previously poured adjacent cycloids or lifted out by crane to be placed in another location. The forms were partially consumed in the building process by the reuse of two layers of 3/8″ marine plywood form sheathing as permanent roof decking. Curved ribs of 2″ × 4″ lumber kept the forms from spreading during pouring operations and at the same time gauged the concrete shell thickness. These ribs remained permanently attached to the curved cycloid shell surface as both spacers for the 1½″ fiberglass roof insulation and nailers for the sheathing.

After the forms were in place, careful and accurate placing and finishing of the concrete was required to meet the exposed line and finish requirements of the Kahn design. Each of the 104′ long roof shells contains six 20-strand steel tendons, each of which was stressed with 140,000 pounds of tension by hydraulic jacks after the 5,000 psi concrete had reached 75 per cent of its design strength. All tendons were factory-engineered and fabricated complete with anchorage plates.

During the development period, Byrne constructed a 12′ long fullsized mockup of the proposed cycloid roof construction to help improvise and improve forming and concrete pouring techniques. This procedure paid for itself several times over, resulting in a more economical cost of the cycloid construction and assuring the quality and accuracy of finished appearance dictated by design specifications. All concrete foundation, floor slab and cycloid roof structural operations were performed by Byrne's own construction crews over a twelve month construction period.

The 46,000 square feet of curved roof surface is sheathed with over 70 tons of 1/16″ thick × 24″ wide sheets of "hard" lead, which contains a high calcium content to increase its tensile strength. The sheets were cleated through a thirty-pound roofing felt to the wood sheathing referred to earlier, with flat locked seams laid out in a symmetrical pattern designed by the architects.

The roofing work was performed from a small sheet metal shop set up on the low soffit flat surfaces of the roof to expedite installation. Originally, lead roofing was desired, but during the estimating period an alternate lead-coated copper material was chosen to reduce the cost. During the eight month design de-

8. Aerial view of the project as of January 1971, showing the roof in varying stages of completion.

9. View of the large cranes used to lift concrete into position for roof pours.

10. Pouring the cycloid roof vaults. The concrete was poured in one foot lifts, using as many as 600 small exterior forms per shell.

velopment period, however, the price of copper increased about 50 per cent while lead prices remained fairly stable. So the lead originally desired was ultimately obtained at a savings under the approved budget.

The structural composition of the main gallery floor was a unique construction system in itself. It consisted of a combination "waffle and two-way flat slab," 14″ thick, which had sandwiched through its 40,000 square foot area approximately twenty-four-thousand 8″ × 30″ × 30″ expanded polystyrene bead blocks. The difference in weight between concrete and these six-pound blocks resulted in a one and a half million-pound weight reduction of the total structural floor system.

The concrete beams within the 14″ floor slab depth were strengthened with a total of sixty-six 34-strand steel tendons that were placed in the forms prior to the pouring of concrete; afterwards, when the concrete had reached 75 per cent of its design strength, each tendon was stretched with 250,000 pounds of force by hydraulic jacks. This minimized slab deflection and allowed the underside of the slab exposed in the office, lobby, and shop areas below to be a continuous flat surface without drop beam supports.

This type of construction was another "first" of its kind to be built in the Fort Worth area. It is a good example of the varied types of innovations employed in the design of the building by the Kahn team that had to executed in a precise, timely, and economical manner by the construction team to meet all the requirements of the owner.

The structural concrete material exposed in portions of the building was selected after many trial samples of different colored cement and aggregates were reviewed by the architects, owner, and contractor with the concrete supplier. A soft light gray color was required to afford a pleasing blend where it came in contact with the soft tones of the Roman travertine veneer at both interior and exterior elevations of the building.

To achieve the desired effect, special procedures of concrete placement, form surface treatment, and mixing were required of the contractor to achieve a consistent harmonizing effect. This involved tight "on-the-job" and "mixing plant" control techniques to assure overall consistency during the thirty months of pouring the exposed concrete.

A total of 12,000 cubic yards of ready-mix concrete weighing 25,000 tons, along with 730 tons of mild reinforcing steel, 42 tons of post-tensioning tendons, 200,000 square feet of wire mesh, and 417,000 square feet of concrete forms were required for the project.

Portions of the building's interior and exterior walls, gallery floors, portico porches, and entrance steps were sheathed with over 1,000,000 pounds of soft-toned Roman travertine shipped from Italy in

11. Close-up of formwork at the end of a roof shell.

12. South wall of the building under construction as seen from the form fabrication area.

13. View of the library mezzanine floor under construction, June 1971, showing steel reinforcing and polystyrene inserts in place.

14. Interior view of completed gallery space.

15. View of conservator's area and courtyard.

seventeen boatloads over a period ranging from fall of 1971 through spring of 1972. Also, 83,000 pounds of granite, 30,000 square feet of brick pavers, and 1,000 square feet of slate flooring was required for additional interior and exterior finishes. This work required accurate setting and close sequence scheduling of delivery of material.

56,000 square feet of washed exposed aggregate topping requiring 500 tons of "dark cedar" crushed marble aggregate was installed for the staff parking areas, pool bottoms, and walks. Travertine and exposed aggregate joints were sealed with 18,000 linear feet of silicone sealant.

Approximately thirteen tons of mill finish stainless steel were fabricated for doors, frames, wall panels, and glass-setting members. These items required a high degree of workmanship and accuracy in manufacturing to achieve the effect desired by architect Kahn. The stainless steel work was shop-fabricated, dismantled, crated, and shipped to the job, where it was installed by local workmen supervised by the fabricator and Byrne.

A custom-fabricated, motor-operated stainless steel-clad rolling door 14′ high × 17′ wide and weighing over 6,000 pounds was manufactured for the receiving dock. All kitchen area, lounge and darkroom cabinets were also custom-made of mill finish stainless steel. Another 35,000 pounds of various custom ornamental metal work was also fabricated.

A continuous skylight system, made of extruded aluminum and double layers of clear "plexiglas" panels and centered along the crown of the cycloid roof shells, was developed to allow filtered sunlight into the gallery areas. 1,200 linear feet of these custom-manufactured units were installed, requiring close contractor coordination regarding construction accuracy in relation to roof shell construction, lead roofing, the aluminum reflectors beneath the skylights, and the museum security program.

Director Richard Brown's and architect Kahn's demands for a consistent light pattern throughout the galleries resulted in the creation of unique continuous inverted gull-wing shaped aluminum reflectors hung beneath the skylights, which distribute light evenly across the downward curving surfaces of the concrete roof shells. In addition to their reflective features, these reflectors provide security features at the skylights as well as continuous light tracks for special illumination of the art on display.

One hundred sixty-eight of these 6′ to 10′ long units averaging 250 pounds each and totaling 1,200 linear feet were fabricated after months of studying the effects of five different mockups. Thousands of closely spaced pin-point holes in each reflector emit sunlight directly into the galleries without it directly touching the paintings on display. These units were fabricated on special jigs developed for the project to meet the accurate tolerances required, and Byrne's own forces connected them directly to anchors em-

16. Exterior view of completed building.

17. View of upper entrance area with reflecting pool in foreground.

bedded in the cycloid shells one year earlier.

The finish millwork and wood floor requirements resulted in a nationwide search for approximately 50,000 square feet of select quarter-sawn white oak of varying thicknesses. After finalizing details, it took over a year to accumulate the necessary materials. Twenty thousand square feet of 1″ thick gallery flooring was installed in a detailed and custom pattern. Thirty thousand square feet of combined millwork, doors, paneling, trim, and cabinets were also fabricated for the project. These items required precision installation within surrounding travertine, stainless steel, and concrete surfaces.

The glass and glazing requirements were also complex. Twelve hundred linear feet of 1/2″ thick strip glazing were installed along the perimeter of the exterior portions of the cycloids. These were embedded in stainless steel channels set in the concrete cycloid structure and glazed with silicone sealant. Thirty-eight hundred square feet of 1″ insulated environmental glass were installed at the upper gallery level to reduce heat loss and eliminate condensation, and special tempered glass was installed in the conservation areas.

The mechanical and electrical requirements are too complex to adequately convey in the space allowed here; however, the sophisticated design requirements dictated a close coordination between Byrne, the mechanical and electrical subcontractors, and the mechanical engineers.

Basically, 500 tons of air conditioning and 4,000,000 BTU's of heat were required. Summer air conditioning is provided by chilling the air temperature to 40° to dehumidify it and heating it up to 72° room temperature. Winter air is heated with moisture added to maintain a constant 45–50 per cent humidity to eliminate deteriorating effects on the art. The controls for this system were custom-designed to maintain the building temperature and humidity requirements and were integrated into the building security system.

A total of 23,000 linear feet of all sizes of piping and 234,000 pounds of sheetmetal duckwork were pre-ordered early in the project for price protection before final mechanical design requirements were known.

The water in the west entrance reflecting pools is recirculated with two 40 horsepower, 3,500 gallon-per-minute pumps to produce a continuous spilling water effect over the upper pool walls.

The four-cell, 500-ton capacity architectural concrete cooling tower was custom-designed and job-built exclusively for the unique Kimbell mechanical requirements.

The building has a 4,000 amp electrical service, which is normally enough for a ten-story building, 100 miles of copper wire, and over 10 miles of electrical conduit and wireways. The facility also has its own self-contained 150 kilowatt emergency generator system, cooled by the reflecting pool water, that automatically will energize any selected electrical circuits in case of power failure.

Thos. S. Byrne, Inc. is particularly proud of its safety record on the Kimbell project that lasted 1,130 calendar days. Byrne's own job crews spent 208,000 man-hours and its subcontractors 130,000 man-hours in constructing the unusual building and only experienced two lost time accidents of minor nature throughout the three year construction period.

There is no doubt that the Kimbell Museum would not exist in its present form without the contributions of Dr. August Komendant, the structural designer engaged in the project by architect Kahn. His innovative solutions to the demands of both architect and owner for vast spaces free of columns and for openings in the crowns of the cycloidal shaped roof shells would probably not have been developed by a less talented structural designer.

As in any building project, the importance of communicating the designer's ideas into workable

18. Detail of entrance portico and side wall.

and realistic plans and specifications cannot be understated. This task fell to the associated architects, Preston M. Geren, Architect & Engineer & Associates, who contributed significantly to the smooth and orderly execution of the work.

Kimbell's building manager was retained early in the project to help determine and develop with the architects the complex technical design requirements pertaining to mechanical, electrical, and security systems. This proved to be invaluable in terms of promptly clarifying and resolving questions of complex nature throughout the work period.

The services of the general contractor played an important part in helping the owner obtain the Kahn design within a realistic budget, and the result clearly underlines how three vital ingredients, quality, time, and cost, can be blended together to produce an outstanding structure.

In terms of cost, the building was around $60.00 per square foot including sitework, which was unusually modest in terms of quality museum construction.

In terms of time, museum director Brown and architect Kahn did not allow lack of time to hasten significant design decisions that they considered important. In effect, time was "bought" to allow a sufficient period to find the proper solution to all requirements.

In terms of quality, the building exhibits its own quality. In a clear, direct manner the museum displays not only the artwork it contains, but also the thought, perseverance, and persistence that both the designers and builders put into it.

NOTES

Owner

Kimbell Art Foundation
Museum director: Richard F. Brown
On-site representative: E. B. Brown

Architect

Louis I. Kahn, FAIA, architect
Project architect: Marshall D. Meyers

Associate architect: Preston M. Geren & Associates
Project coordinator: Frank H. Sherwood

Consultants

Structure: August E. Komendant
Lighting: Richard Kelly; Edison Price
Acoustics: C. P. Boner and Associates
Landscape: George Patton
Graphics and exhibition design: Laurence Channing
Electronics: Thomas Electronics

Construction

Thos. S. Byrne, Inc., general contractor
Project manager: A. T. Seymour, III
Job supervisor: Virgil Earp

Landscape: Vernon S. Swanson

The following is a listing of subcontractors, material suppliers, and manufacturers who contributed to aspects of the project discussed in the article.

Excavation: Joe R. Starks Construction Company, Fort Worth, Texas.

Steel tendons: Prescon Corporation, San Antonio, Texas, fabricator.

Lead roofing: Murdock Lead Company, Dallas, Texas, supplier; installed by McClung Roofing Company, Fort Worth.

Polystyrene bead blocks: United Floatation Industries, Irving, Texas, manufacturer.

Concrete: Texas Industries, Inc., Fort Worth, supplier.

Marble, granite, brick pavers, slate flooring: McHarg Marble and Tile Company, Fort Worth, supply and installation.

Crushed marble aggregate: Cactus Canyon Quarries, Marble Falls, Texas, supplier; installed by Couch Terrazzo Company, Fort Worth.

Silicone sealant: General Electric Company, supplier; installed by Max Eubank Roofing Company, Fort Worth.

Stainless steel doors, frames, wall panels, and glass setting members: Louis Hoffman Company, Milwaukee, Wisconsin, fabrication and installation.

Receiving dock door and miscellaneous ornamental metal work: Southern Industrial Steel Company, Arlington, Texas, fabricator.

Cabinets and kitchen equipment: Vinson Manufacturing Company, Dallas, Texas, manufacturer.

Skylights: Naturalite, Inc., Dallas, Texas, manufacturer; installed by McClung Roofing Company, Fort Worth.

Aluminum reflectors: Offenhauser Company, Houston, Texas, fabricator, in cooperation with Louis Hoffman Company, Milwaukee, Wisconsin.

Millwork, doors, paneling, trim, and cabinets: Robert Shaw Manufacturing Company, Fort Worth, fabricator.

Glass and glazing: PPG Industries, Inc., Fort Worth.

Mechanical: General Engineering Corporation, Fort Worth, sub-contractor.

Electrical: Shotts Electric Company, Fort Worth, sub-contractor.

19. Aerial view of completed building, with landscaping nearing completion.

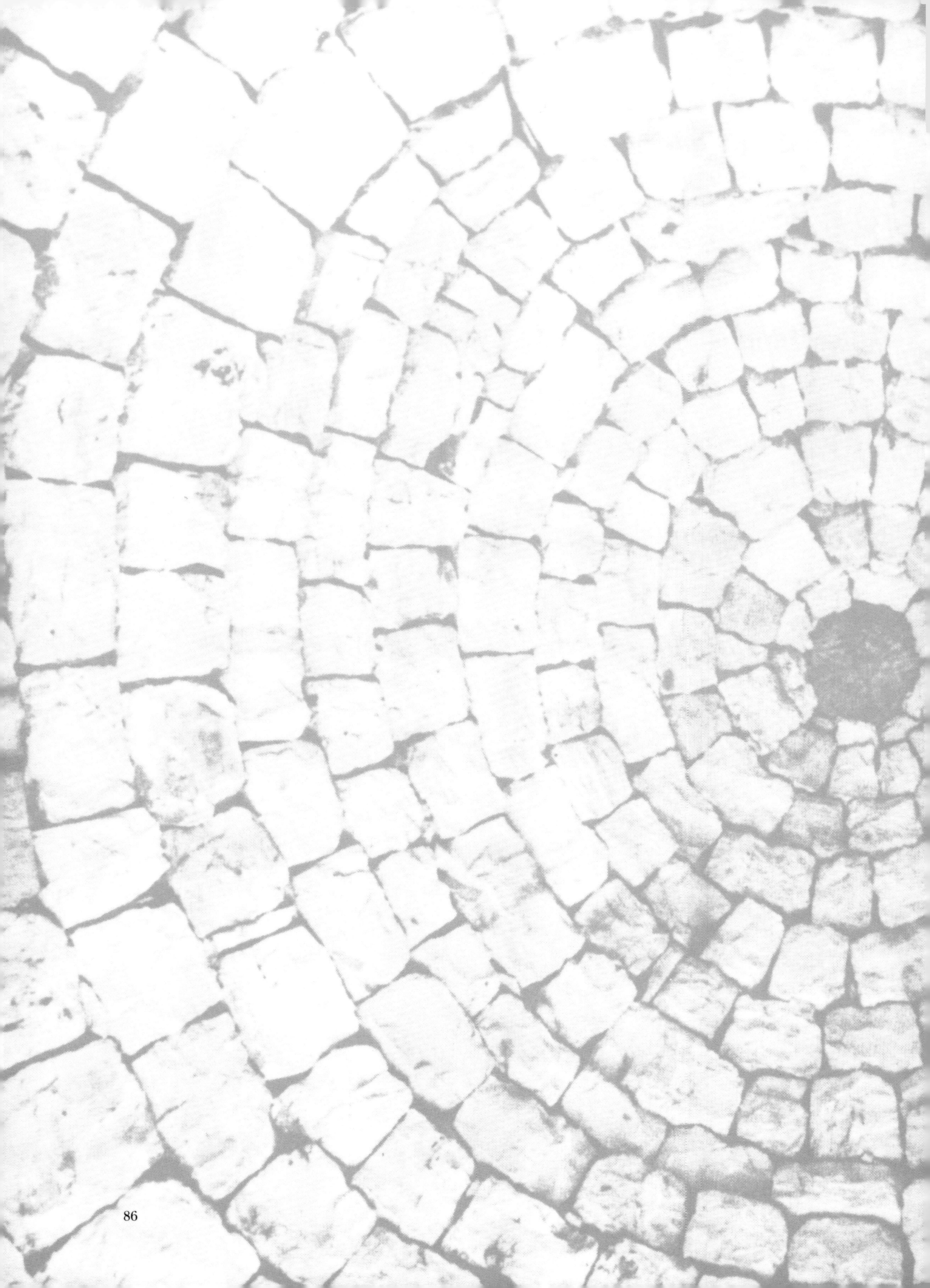

FABULA RASA

Emilio Ambasz

THE LITTLE VILLAGE was in the grip of fear, fear of Divine rages and of human passions. One of the men started to build a construction, circular in plan, cylindrical in volume, and with a domelike roof. He used stones, wood, and mud. His travails finished, he came back to tell the group the building he had erected was in the shape of the Universe, and inside dwelt the Universe's gods; he called it *a temple*.

Then, using a rod he had taken from the temple, he made a circle around the village, and with the help of others he encircled it with a high wall built of earth and stones. In the center of the village, next to the temple, he erected a large hut, which he then covered completely, except for the entrance, with a mound of earth. On top of this mound he placed vertically six large stone slabs. That he called *his home*. The others called it *the palace*.

When he died, his body was laid down inside the hut he had called his house, together with all his belongings, and his son covered the entrance with the large stone slabs he removed from the mound's top.

People have said this was how architecture began.

1. The roof of the audience hall in the Imperial Palace in Beijing.

YINGZAO FASHI: CHINESE BUILDING STANDARDS IN THE SONG DYNASTY

Else Glahn

IN 1918, a manuscript copy of a book from the Song dynasty (960–1279) was discovered in a library in Nanjing, China. The book was called *Yingzao fashi* or *Building Standards*. Printed for the first time in 1103, it describes building technology, preliminary calculations, foundations, carpentry, joinery, and manufacture of a number of building materials such as bricks, tiles, paints, mortar and plaster. It continues with calculations of manpower and materials and ends with drawings of constructions, types of buildings, and decorations.

In China, architecture was traditionally regarded, like sculpture and ceramics, as an anonymous craft. Public buildings were erected by craftsmen employed by the government. There were no architects; the technology of building was handed down from master to apprentice. In contrast, the Fine Arts, painting and calligraphy, were cultivated by the scholar officials. Because modern Chinese historians had been out of touch with architecture, they regarded it as unchangeable and unchanged through history. Recent research based on the *Building Standards* has proved, however, that a number of changes have actually taken place.

The *Building Standards* was compiled during the years 1097 to 1103 by imperial order to the Office of Construction, a department under the Ministry of Work. It was a revised edition of a thirty-year-old work that has disappeared completely. The editor, Li Jie, was vice-director of the Office of Construction. His introduction states that he "questioned craftsmen who explained everything." In February of 1103 he received imperial permission to have his work printed.

The *Building Standards* was compiled during a period when the Chinese government changed its economic system considerably. Until the 1060's most taxes and expenditures were paid in kind, and public work was carried out as a labor duty. After this time, most work was paid for in cash, and the craftsmen received wages. This change was caused by a need for close control of government expenses. The *Building Standards* gives methods for the exact calculations of labor and material necessary for all sorts of building construction, down to the smallest details. The introductory chapters about building technology are necessary prerequisities for the calculations; they were not meant to teach officials, responsible only for accounts, how to become architects, nor were they meant to teach construction to the craftsmen. But today the *Building Standards* can guide our research in Song architecture.

Most remarkable is the high level of standardization described in the work. It is so thorough that it implies a long preceding period of development. Western systems of standardization in building technology are comparatively new. Archaeological finds have proven that when this system was written down shortly before the year 1100 it had existed for more than a thousand years, so the traditional Chinese system of standardization is about two thousand years old. The Chinese system of standardization did not originate with an official at a writing desk. It developed organically through the practical experience of the carpenter.

The Chinese system of standardization in building technology has "unit as basis": measures are given in units and sections, not in feet and inches. Buildings used different sizes of unit according to their relative importance. Once unit size was established, the carpenters would use a unit rule instead of a foot rule.

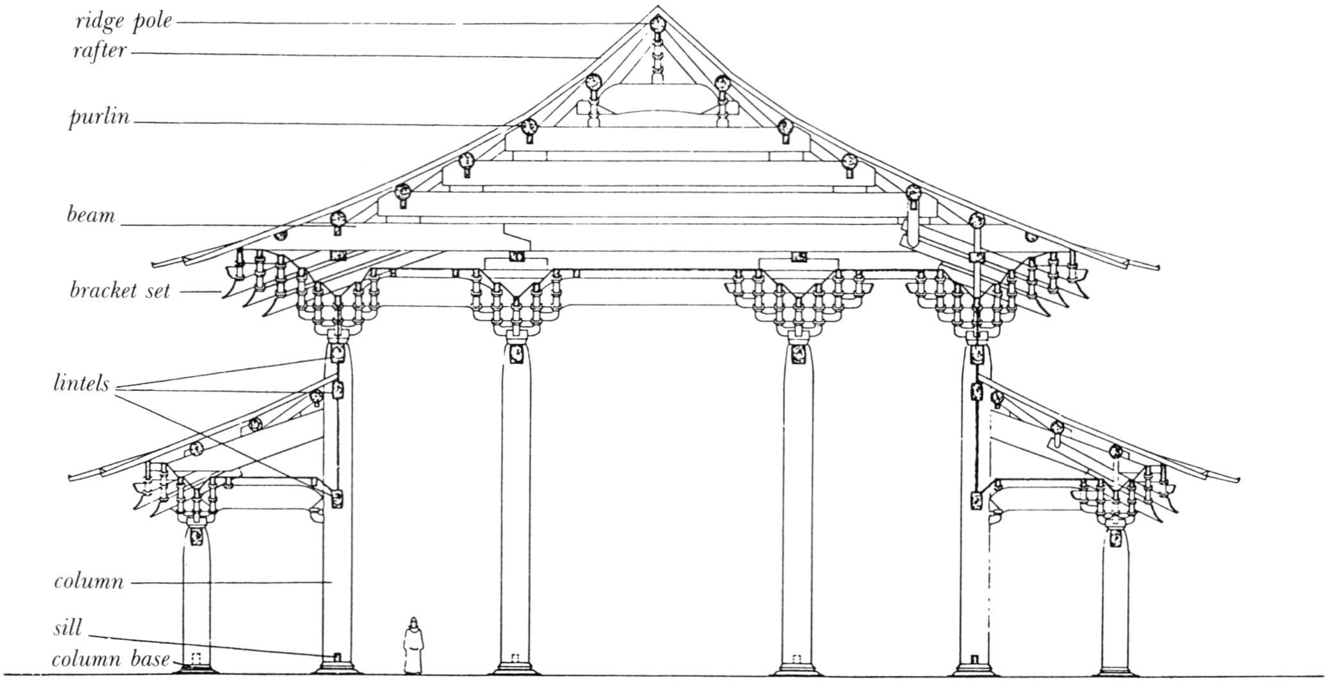

2. *Cross section of a big hall according to the* Building Standards.

The same flexible system of construction was used for a wide range of buildings, from modest storehouse to large palace.

The first impression made by a typical Chinese building of this period comes from its curved overhanging roof, which is heavy but seems to float weightlessly in the air (Fig. 1). It is carried by wooden columns standing on stone bases, and fixed at the base by wooden sills (Fig. 2). At the top they are connected lengthwise by lintels and crosswise by beams that are piled up one above the other, each one shorter than the one below. The curve of the roof determines the length of the upper beams. In the roof construction the long purlins, going from one end of the building to the other, rest on the ends of the upper beams (Fig. 2). The uppermost purlin is the ridgepole. Short rafters placed as densely as teeth in a comb span from one purlin to the next. This is the main difference between a Chinese and a Western roof, in which the rafters go uninterrupted the whole way from the ridgepole to the eaves. This construction makes the Chinese roof frame far more flexible than the Western one. The rafters are covered by a wooden sheathing on which the mission tiles are placed in mortar. The weight of the roof is transferred to the columns by projecting superimposed bracket arms placed wherever a beam and a column meet. These complex bracket clusters, which carry the deep eaves, form a ribbon around the building.

All walls are nonbearing curtain walls made by wattle and daub or sun-dried bricks. They are placed according to need for protection and division, not according to demand for support. In main buildings both ends and often the back side have no openings, but are filled out completely. The front aspect, facing south, is open, with windows and doors made of wooden lattice. The entrance is in the central span; therefore, the number of spans is uneven. Though fragile and vulnerable to the climate, the walls are protected by the overhanging roof, which keeps the summer sun out and lets the winter sun in.

In Chinese architecture space is as important as physical mass; the interaction between a courtyard and the buildings lining it is significant. A building is rarely isolated and detached. A traditional Chinese building is situated in a rectangular compound and is subordinate to the general layout of this compound (Fig. 3). The compound is surrounded on all four sides by a wall that is oriented according to the cardinal directions, with the main buildings and the gate placed on a south-north axis, north being the direction of superiority.

This description fits the public building all over China for more than a thousand years. Domestic houses are still usually built of local material, but public buildings were timber framed, with curved tile roofs. This construction method proved adaptable to the different climates in this big country, from the

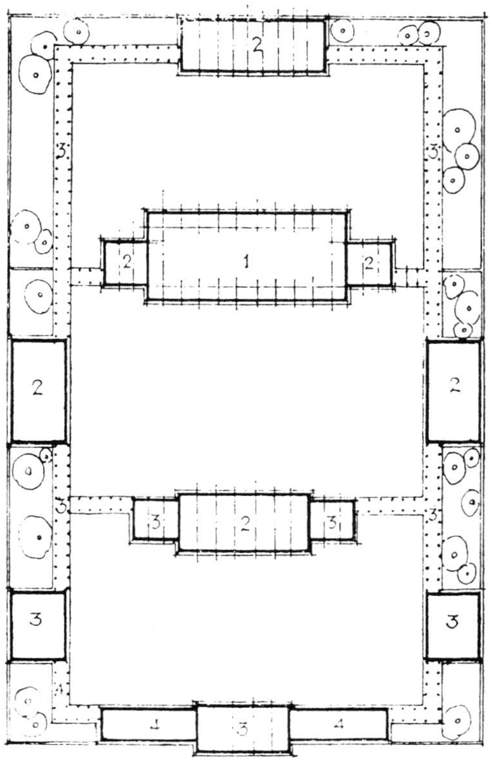

3. *A building compound.*

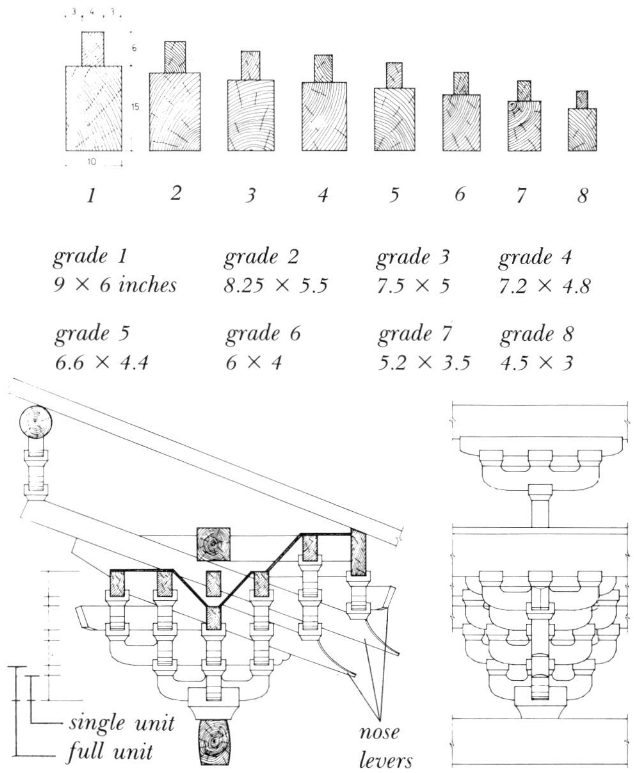

4. *Standard dimensions according to the* Building Standards.

Manchurian winters in the north to the subtropic provinces in South China, and from the deserts in Central Asia to the coast of the Chinese Sea.

The Chinese system has the cross section of a construction member as its unit, while the Western has space, the length of a span or a smaller opening, as its unit. The aim of both systems is a rationalization of the building process, but there are other implications to the Chinese system.

In the Song period buildings were divided into eight groups according to their size, and for each group a standard unit of timber was used. The timber units were used for all sorts of construction work in timber as well as in other materials. The ratio between the depth and width of each timber unit was 3 : 2. The depth and the width were then subdivided into 15 : 10 and one fifteenth of the depth was called one section (Fig. 4). The *Building Standards* says:

> All methods for timber construction have timber units for their base. Timber units have eight grades of measure, each used according to the size of the building:
> First grade: the depth is nine [Song] inches and the width six inches. 0.6 inches makes one section. This grade is used for halls nine to eleven bays long. Second grade: the depth is 8.25 inches and the width 5.5 inches. 0.55 inches make one section. This grade is used for halls five to seven bays long . . .

The third grade is 7.5 × 5 inches, the fourth, 7.2 × 4.8 inches, the fifth, 6.6 × 4.4 inches, and sixth, 6 × 4 inches.

> . . . Seventh grade: the depth is 5.2 inches and the width 3.5 inches. 0.35 inches makes one section. This grade is used for small buildings or pavilions. Eighth grade: the depth is 4.5 inches, the width 3 inches. 0.3 inches makes one section. This grade is used for canopy ceilings or pavilions with complicated bracket sets.

One Song inch was approximately 3.2 cm. This means that the timber unit of the first grade corresponds to approximately 28.8 × 19.2 cm, with a section of 1.92 cm, and so on down to the eighth grade, which has a depth of approximately 14.4 cm and a width of approximately 9.6 cm, with a section of approximately 0.96 cm.

The unit was derived from the cross section of the bracket arm. A unit could be a measure of length as well as a measure of area. As a measure of length it was fifteen sections, the depth of a bracket arm. A rectangular piece of timber could be placed on top of an arm to strengthen it. Such a stiffener was six sections deep and four sections wide (Figs. 4 and 5). The depth was the same as the distance between two superimposed bracket arms. In some cases the expression "full unit" was used for the length of twenty-one sections or for the depth of an arm plus a stiffener.

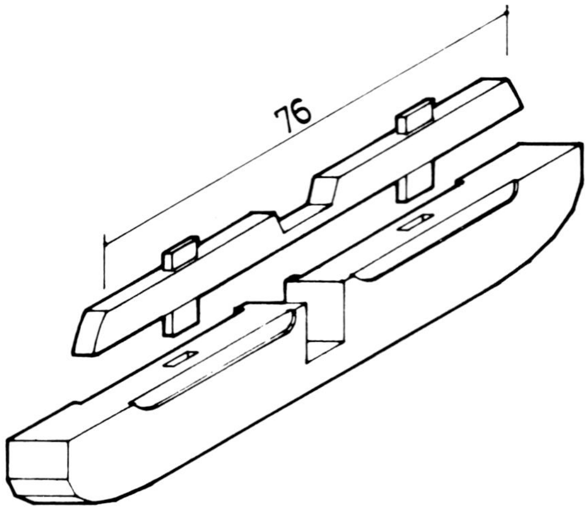

5. *A bracket arm with a stiffener.*

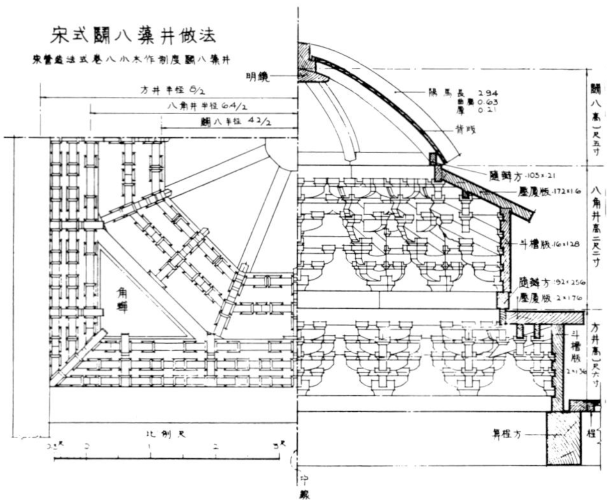

6. *A canopy ceiling.*

7. *Levelling of a foundation.*

Throughout the *Building Standards* almost all measures are given in units and sections, not in feet and inches. A unit was fifteen sections; if not, "full unit" was explicitly stated. By giving proportional measures, the descriptions suited buildings of different sorts and sizes. The craftsmen did not need inch rules, for they used unit rules instead.

A building's grade of unit was determined by the number of spans in its front aspect. A major hall could be surrounded by a colonnade. In this case the colonnade was built of a unit one grade smaller than that of the hall (Fig. 2). In a ceremonial hall the ceiling over the seat of honor could be shaped like a cupola as a canopy for the deity or ruler (Fig. 6). Such a cupola was built in timber of the smallest grade of unit. The smaller grade of colonnade and canopy set off the majestic impression of the hall itself.

In a courtyard the different buildings used different grades of unit to set off the main building, which was usually situated in a second courtyard facing south and placed on the central south-north axis (Fig. 3). If its unit was grade one, then the side buildings would have a unit of grade two, as would the central buildings to the north and south. The side buildings of a hall with a unit grade of two would be of grade three, and the gate or entrance building would also be of grade three. The side buildings to the gate would then use a unit of grade four.

It was, in principle, only necessary for a team of craftsmen to know a building's number of spans in order to construct and erect it. No drawing table was

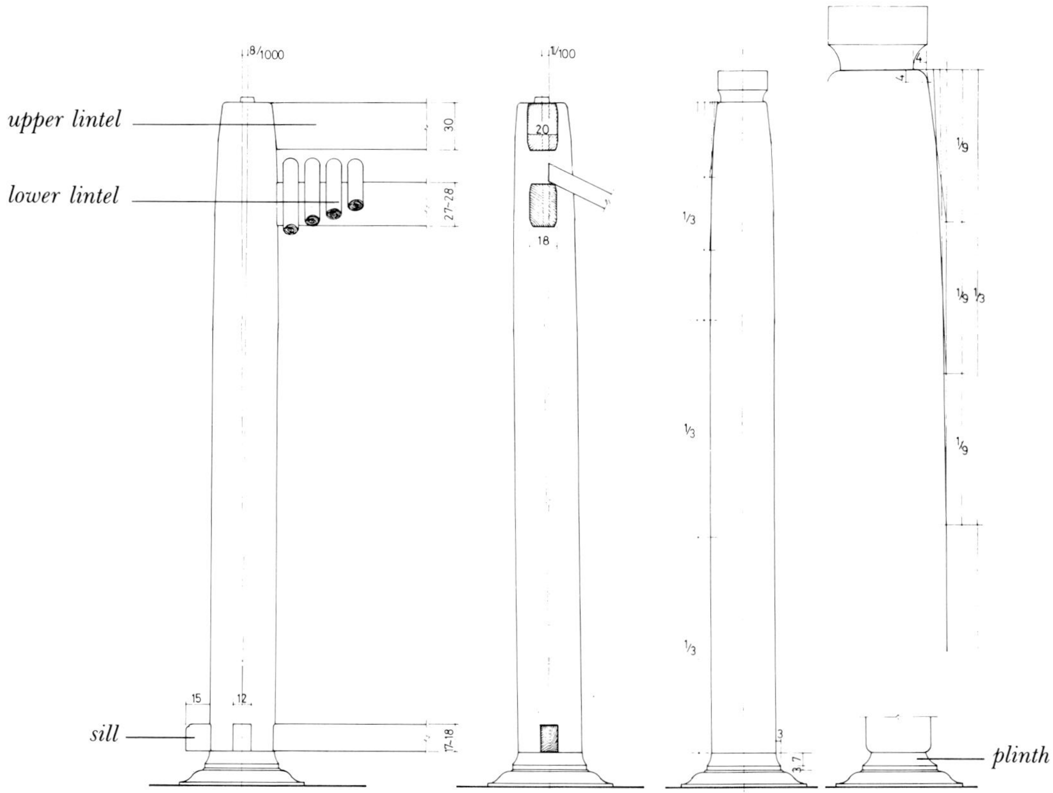

8. Column.

needed. Simple buildings were erected by craftsmen without supervision. But the standardization system was far from rigid. Though the dimensions of the construction members were fixed, the length of spans and the height of columns could vary within certain limits. A span could be between 200 and 375 sections long, and, if necessary, up to 450 sections long. The central span was wider than the side spans. A column could be up to 375 sections long, but it could never exceed the length of the central span. Such variations made it necessary to adjust the number of brackets in the bracket sets and the length of several ties. The erection of complicated buildings with such adjustments was supervised by master craftsmen.

Site Preparation

Before building operations could begin, the site had to be oriented in exact accordance with the cardinal points. This was done in order that the front side, the most important side of the building, would face south. In a palace building the seat of the emperor and in a temple the seat of the deity faced south. When the orientation was correct the site was leveled to be exactly horizontal.

The foundation consisted of layers of rammed earth alternating with layers of broken brick and gravel, also rammed. The earth was beaten with a wooden rammer and the broken brick and gravel crushed with a circular stone swung by two men (Fig. 7).

When the foundation was solid it was covered with stone slabs which were to be approximately 96 × 64 cm and 20 cm deep. The stone bases for the columns were fitted with the stone slabs and had a diameter double the diameter of the column. The surface of each column base was carefully leveled.

Building Layout and Erection

The distance between columns varied. On the front side the central bay was longer than the others. Inside a building the columns were placed independently of the columns in the outer walls, according to the function of the building. In a temple, the deities were placed far back in the hall, with space in front of them for the devotees.

The columns in the outer walls were connected at the base by sills with a depth of one unit and two or three sections, or 17–18 sections (fig. 8). The width of the sill was 2/3 the depth.

Columns

The heads of the columns were connected by lintels two units or thirty sections deep. The width was 2/3 the depth. Where an open colonnade was placed to the front of a building or surrounding it, two or more lintels were placed below the upper one. The roof of the colonnade was carried by one of the lower lintels. Its depth was two or three sections less than that of the upper lintel.

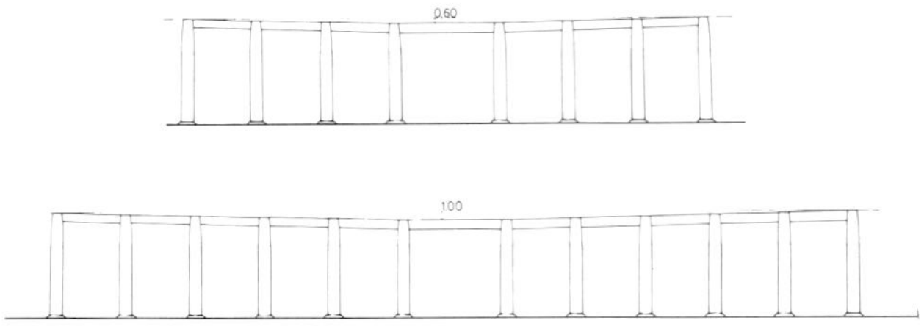

9. Variation in column heights.

10. Building placed on bracket sets. (From the scroll: Quingming shanghe *by Zhang Zeduan, ca. 1117.)*

In a big hall, the diameter of a big column could be as much as 42–45 sections, and in smaller buildings up to two units or thirty sections. Under the columns and on top of the column bases were placed circular wooden plinths, ten sections high. In case of rot the plinth could be replaced without damage to the column. The heads of the columns were shuttle-shaped. This shape was achieved by gradually cutting off the upper third of a column in three segments. The column was cut four sections from the top to fit the bottom of the bearing block. All columns had a slight slant towards the center of the building. In the longitudinal direction the degree of the slant was 1 : 100, and in the other direction it was 1 : 125.

The two columns at each side of the central bay in the front side of a building were shorter than the rest (Fig. 9). The columns were gradually higher toward the corners, and a line through the heads of the columns formed a graceful curve. In a building eleven bays long the corner columns were ten inches or 32 cm higher than the central columns. In a building of seven bays the extended height was eight inches, and so forth. The slant and the different heights of the columns stressed the curve of the roof.

Brackets

The bracket sets on top of the columns constituted very important elements in the frame of the building. So much attention is paid to them in the *Building Standards* that the description of them takes up as much space as the combined descriptions of all the other carpentry members—columns, beams, purlins, etc.

The bracket sets were placed where vertical and horizontal construction members met. They transferred the weight of the roof and the horizontal construction members to the vertical members. They wove the frame together, so to speak.

In the Song period and earlier, many buildings were placed on a series of bracket sets (Fig. 10). A bracket set could consist of four different sorts of structural members: bearing blocks, bracket arms, levers, and a nose (Fig. 11). The blocks supported the arms, the levers, and the nose.

Four sorts of blocks are described in the *Building Standards* (Fig. 12). The biggest was the cap block, placed on top of a column and tenoned into it. The cap block was 32 sections long and wide and 20 sections deep. The other blocks were only ten sections deep. The connection block was placed on top of the lever arm or on the end of an arm extending at right angles to the building. Its top side was 18 × 16 sections. The center block was 16 × 16 sections and was placed at the center of an arm carrying the center of an arm above. Finally, there was the small block placed at the end of an arm parallel to the building. Its top side was 16 × 14 sections.

There were five different sorts of bracket arms (Fig. 13). The petal arm projected at right angles to the building, rested in the cap block, and was 72 sections long. In the upper parts of the sides two eyes were cut out three sections down the side. A petal arm was usually made of full standard, in contrast to the other arms. When the three sections of the eyes to both sides of the arm were subtracted from the width, four sections remained. The cross section of this stiffener, added to the top side of the arm, became 6 × 4

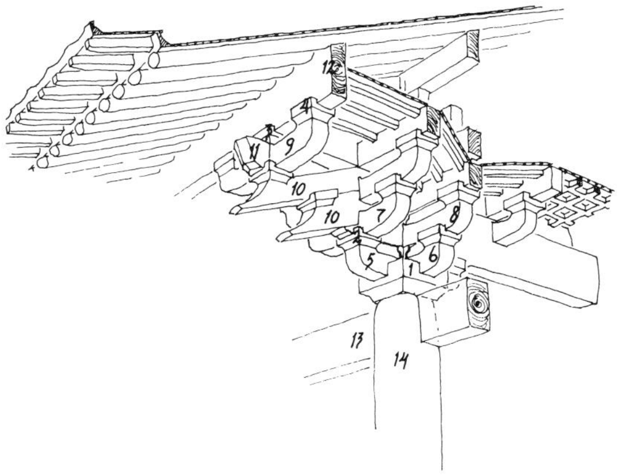

1. *cap block*
2. *connection block*
3. *center block*
4. *small block*
5. *petal arm*
6. *wall arm*
7. *oval arm*
8. *long arm*
9. *regular arm*
10. *lever*
11. *nose*
12. *board*
13. *lintel*
14. *column*

11. *The different members of a bracket set.*

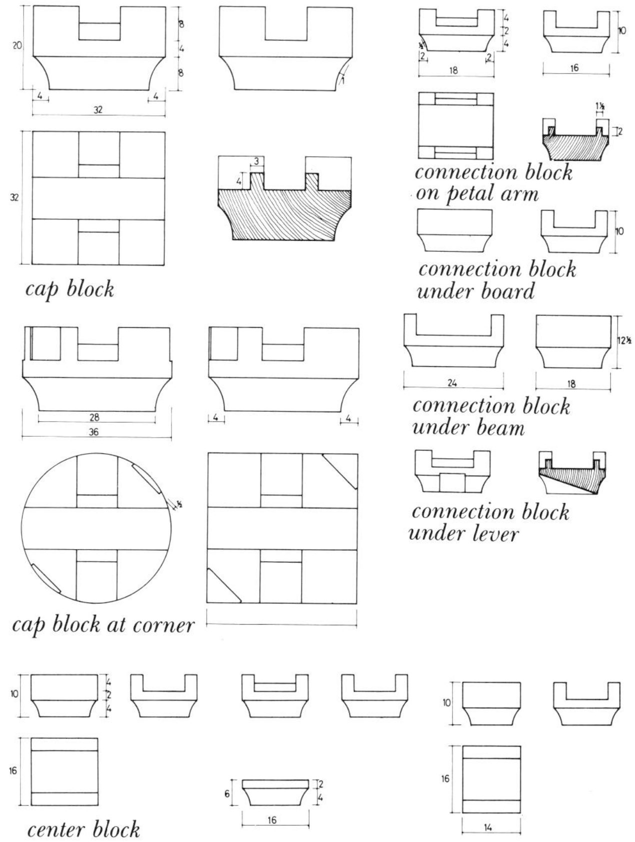

12. *The four bearing blocks.*

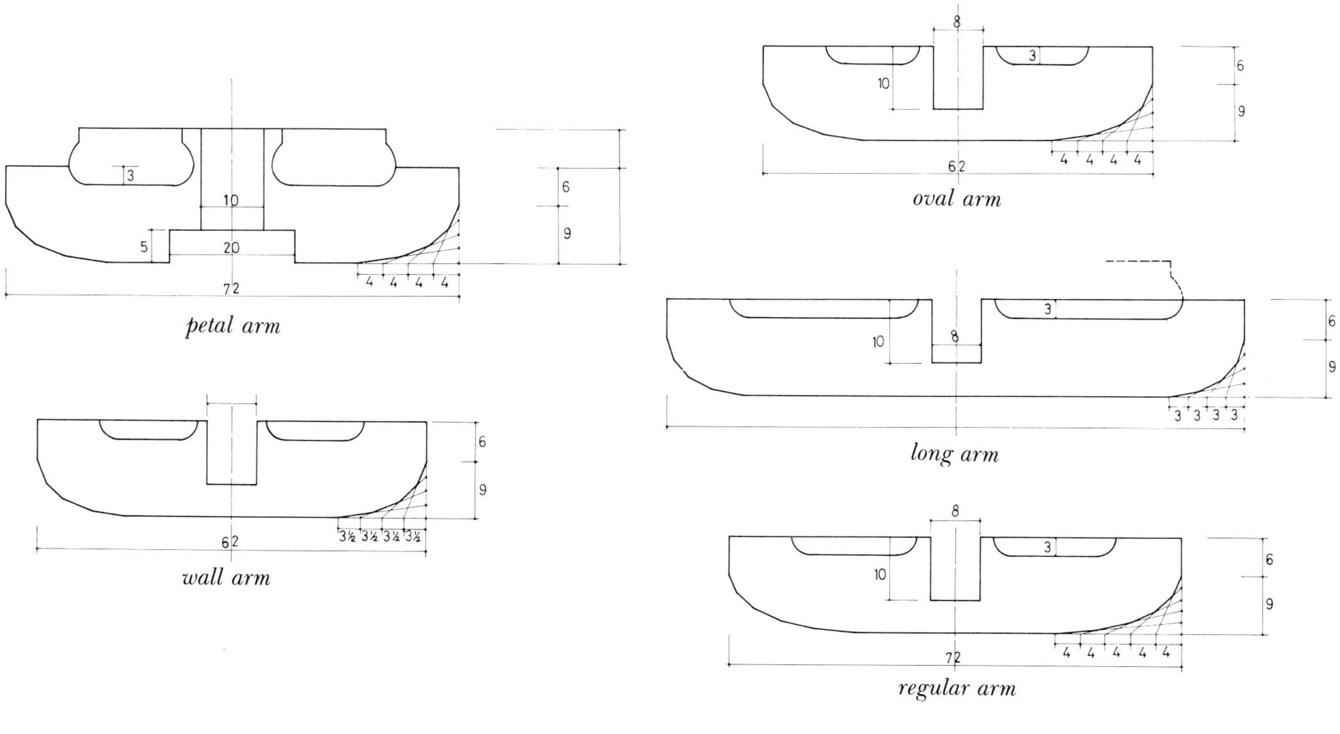

13. *The five bracket arms.*

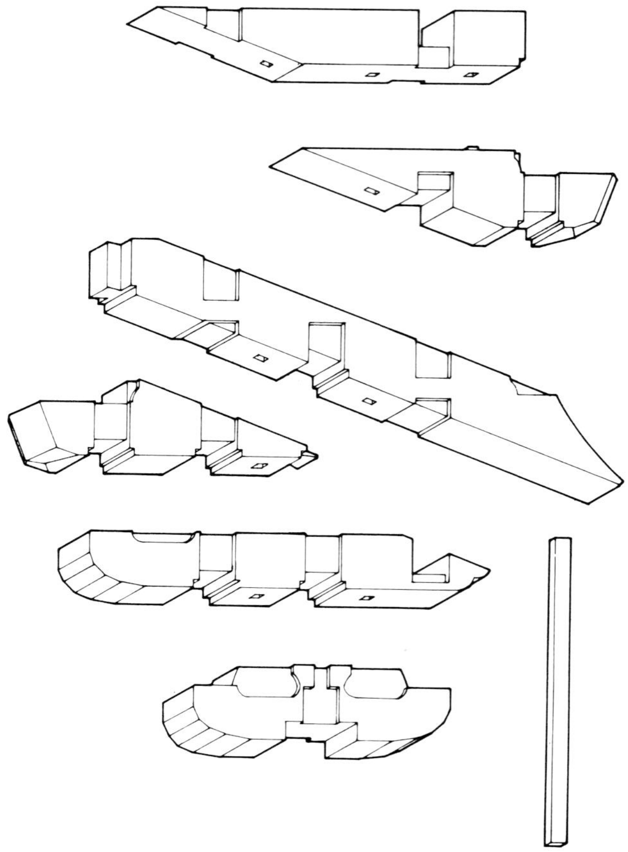

14. Dowel.

sections. These are the measures for the "full unit" given in the chapter about timber units (Fig. 4). The underside and the sides at the center were cut out to be tenoned with the wall arm. Only the petal arm was situated at right angles to the building; the other arms were flush with or parallel to the wall.

The wall arm, which was 62 sections long, was also placed in the cap block flush with the wall. A third arm was the oval arm, parallel to the wall and extending to each side on top of the protruding end of a petal arm. It was 62 sections long and rested in a connection block. There was, further, the long arm in the tier above the oval arm and the wall arm; it was 92 sections long and rested in a center block. The uppermost arm, the regular arm, was 72 sections long. At its center a center block was placed and at each end a small block. The three blocks carried a horizontal board.

The most interesting member of the bracket set was the lever (Fig. 4 and 11). It was a long straight piece of a single unit placed in a block at the center of the bracket set and above the head of a column. One end went down in an oblique line beyond the bracket arms and carried the eaves, while the other end went up in the roof and supported a purlin, acting structurally as a lever arm. Levers were only found in bracket sets carrying eaves. It was this construction member that made the very deep eaves possible.

The nose had no function in a bracket set with levers (Figs. 4 and 11). Originally it was the end of a beam joined with a bracket set, but when there were levers in the set, the nose was only a residual, decorative element.

The members of the bracket sets were all mutually joined and further stabilized by dowels (Fig. 14). A dowel was four sections deep and two sections wide. The length depended on the size of the bracket set. Not a single nail was used in a bracket set.

The number of members in a bracket set varied. A set could consist of a single arm or a single lever with the corresponding transverse arms and blocks. It could also be very complex, with two petal arms and three levers with all the corresponding transverse arms and blocks. The complexity depended on the place of the set. The bracket sets on top of the corner columns were especially complex. The intercolumnar sets were smaller than the sets on top of columns, and the smallest were placed inside a building, where they carried a beam or tie in the roof frame.

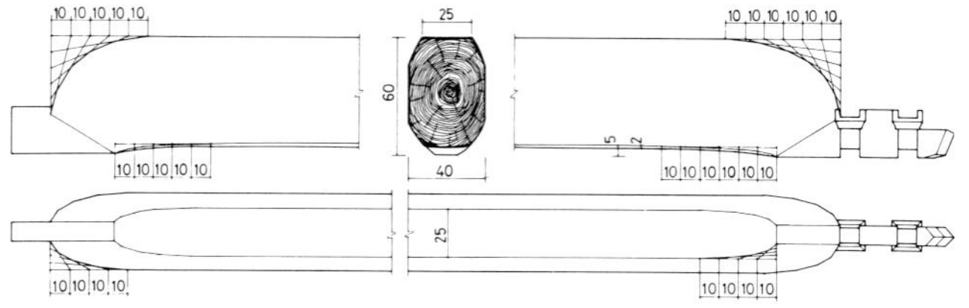

15. Crescent Beam.

Beams and Rafters

Big beams crossed the building, and over them a series of shorter beams were superimposed. Joined with a bracket set on top of a column was the bottommost beam, which was also the biggest of the beams (Fig. 2). In a big hall it could be as much as 60 sections deep, while other beams were two units or 30 sections deep. The width of a beam was always two-thirds the depth.

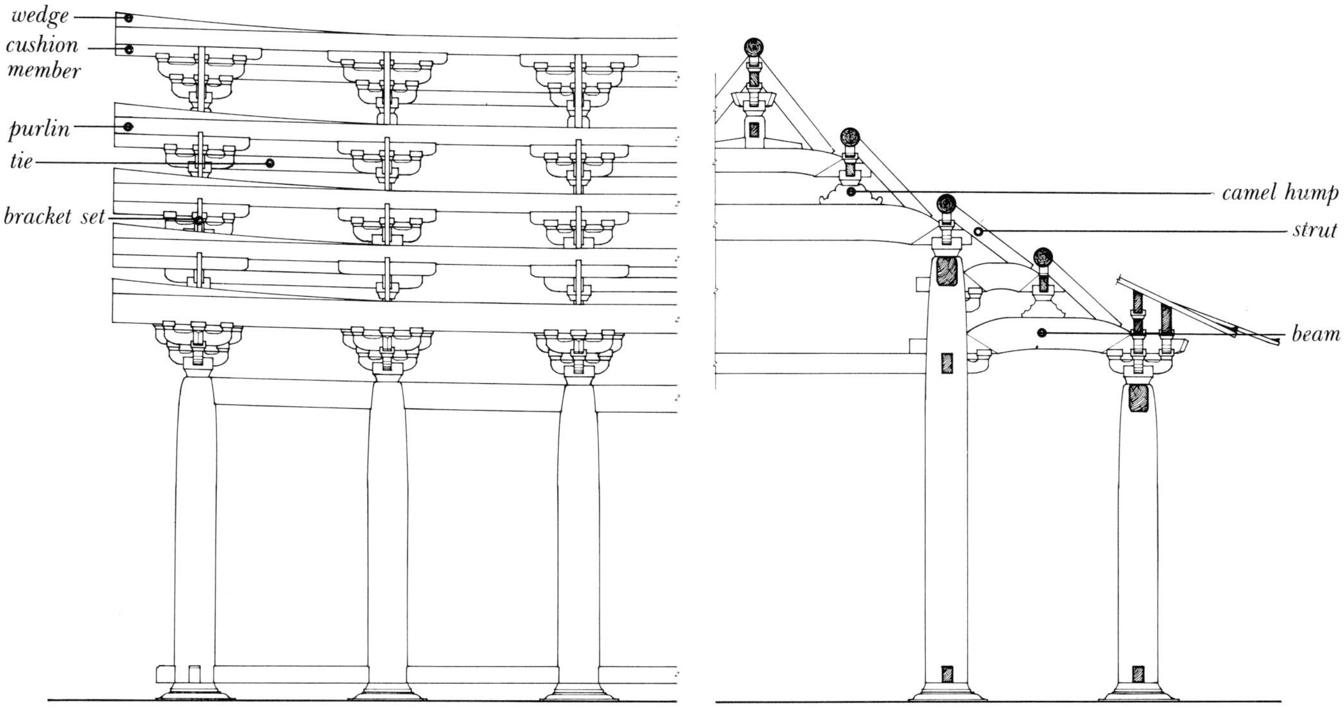

16. Construction members in the roof frame.

In the *Building Standards* is a detailed description of the curvature of so-called moon or crescent beams (Fig. 15). This curvature of the beam was popular in South China, where the capital of the Song dynasty was situated and the *Building Standards* written, but it never became common in North China.

The ends of the superimposed beams rested in the center of simple bracket sets consisting of one or two arms (Fig. 16). These were placed on supporting members, the so-called camel humps, situated near the end of the beam below. The length of the superimposed beams was decided by the curvature of the roof.

In the longitudinal direction, the bracket sets on the camel humps carried the cushion timbers that supported the purlins. The cushion timbers were 10 sections wide and 12 sections deep. The length was 92 to 125 sections. The bracket sets were joined with construction members that tied the spans together, two and two. The length of each tie was one bay, and the cross section, one unit.

At the end of each beam the purlins went from one end of the building to the other. They were circular, with a diameter of 21 to 30 sections in a big hall, and 16 to 17 sections in a smaller building. From the column next to the end of a building a wedge was attached to the top side of a purlin. At one end of the building this wedge was one unit deep. Its function was to form the longitudinal curve of the roof. The purlins were further supported by oblique struts standing on the beam below.

The short rafters were circular (Fig. 17). In big

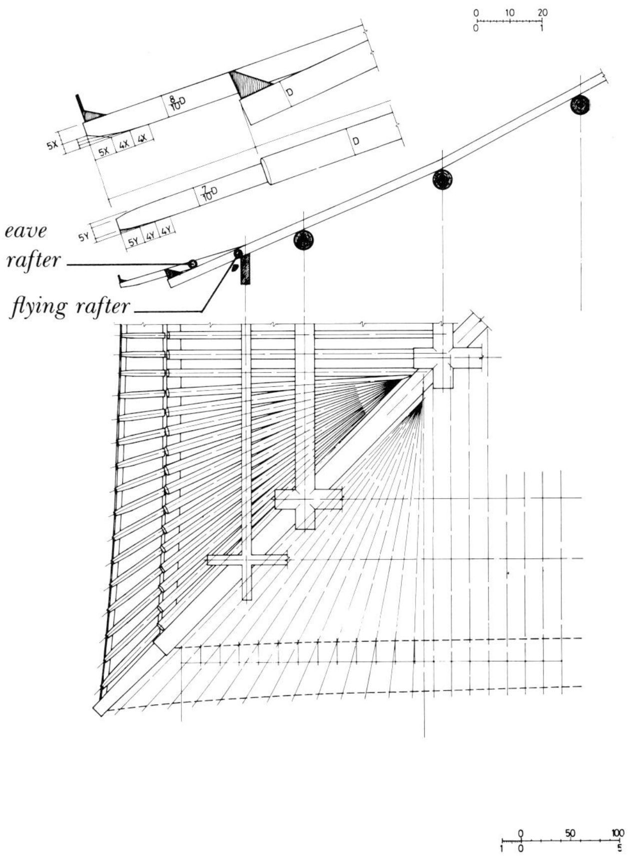

17. Rafters.

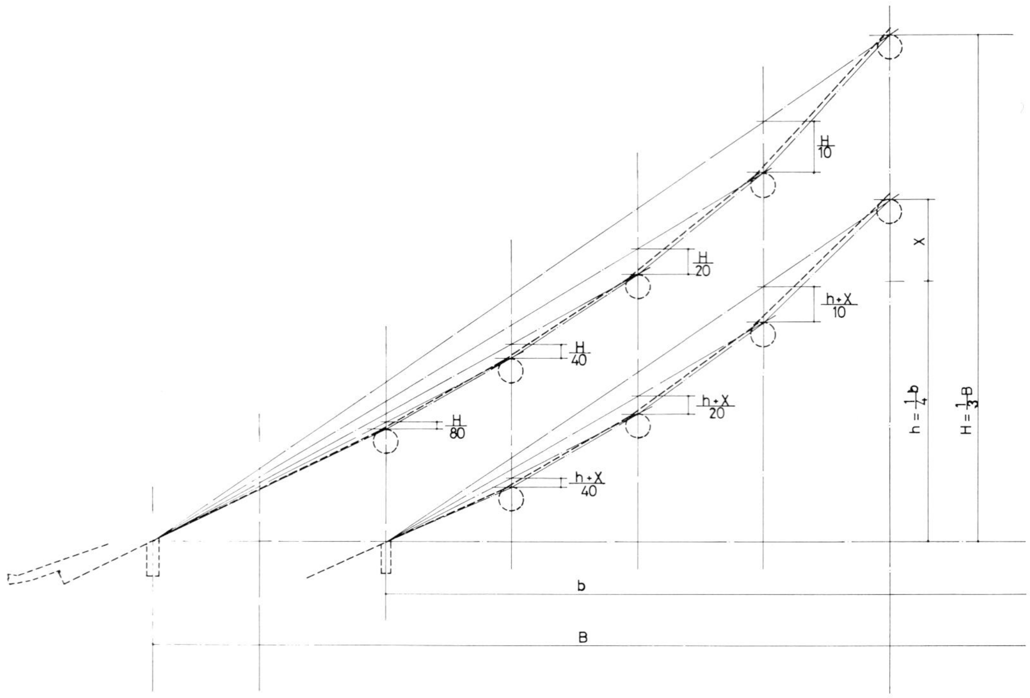

18. Calculation of the roof curve.

halls their diameter was nine to ten sections and in smaller buildings six to seven. The length was the oblique distance between two purlins, and the distance between the rafters was between 9.5 and 7.5 sections. They supported the wings of the bottom tiles. The rafters were nailed to the purlins through the upper end. The other end rested freely on the purlin below.

The eave rafters were longer than the others. The rules for their projection were not derived directly from the size of the building, but indirectly from the diameter of the rafters. The measure was given in feet and inches, not in units and sections. When the diameter of the rafter was three inches, or approximately 9.6 cm, the eave rafter would project 35 inches, or approximately 112 cm measured horizontally. This would be for a building of the sixth grade. When the diameter was five inches, or approximately 15 cm, the eave rafter's projection was 40 to 45 inches, or approximately 128 to 144 cm. This would be for a building of the third grade.

The ends of the eave rafters formed a curve from the center of a building to its corner. In a building five spans long the ends of the eave rafters at the corners protruded seven inches, or approximately 28.4 cm longer than the central rafters. To the top side of the rafters was attached a triangular list.

The so-called "flying eaves" or flying rafters were tenoned into the top side of each of the eave rafters and rested in the list. Their length was 3/5 the horizontal projection of the eave rafters. Another triangular list was attached to the top side of the flying rafters. The flying rafters with their list braked the roof tiles and prevented them from sliding.

The rafters were arranged carefully so that the center of the distance between two rafters was in the vertical line of the center of the bracket set below. At the corners the rafters were spread like ribs in a fan. From the purlin next to the eave purlin the upper ends of the rafters were cut thin and placed so that the eave rafter next to the hip would have turned almost 45 degrees.

Roof

The construction of the roof curve was called "elevation and depression." The elevation was the vertical distance between the ridge pole and the eave purlin (Fig. 18). The depression was the vertical distance between the position of a purlin if it were situated on a straight line from the ridge pole to the eave purlin and its actual position in the roof curve.

If the distance between the eave purlins across a big building is called B and the distance of the ridge pole from the line between the eaves called H, the H will be $B/3$. A line from the ridge pole to the eave purlin was divided into as many sections as there were purlins. The depression of the first purlin from the ridge would be $H/10$. A new line was drawn from the actual position of this purlin to the eave purlin and

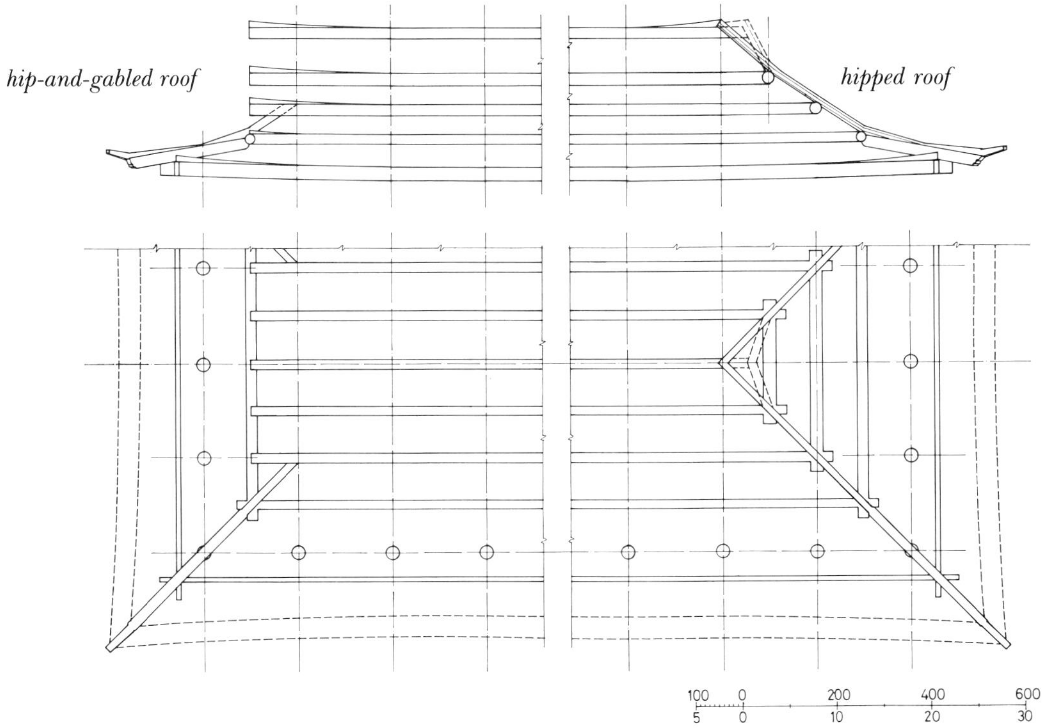

19. Hip-and-gable and gable roofs.

divided into as many sections as there were purlins left. The next purlin was placed $H/20$ below this line. The following purlin was placed $H/40$ under a line from the base of the second purlin to the ridge, and so on.

The rules for the roof curve of smaller buildings were slightly different. A line was drawn between the eave purlins and the length of the line was called b. The distance of the ridge pole from this line was h, and h was $b/4$. In the calculation of the depression a factor x was involved. This x was between $8b/100$ and $3b/100$, according to the size of the building. The depression of the first purlin was $(h + x)/10$, the next was $(h + x)/20$, and the next, $(h + x)/40$.

Ceremonial halls and other important buildings had hipped roofs. Less important buildings had hip-and-gabled roofs. Only the simple houses of the common people had gables. In a hipped roof not only was the line of the ridge curved like the purlins, but the ends of the ridge were further extended three feet, or approximately 96 cm (Fig. 19).

A building could have a ceiling inside or the roof construction could be open. A ceiling was made of timber and was checkered or coffered. A ceremonial hall could have a cupola or canopy ceiling over the deity or the throne.

Examples

To what extent the *Building Standards* describes construction methods current in the Song period can be seen by comparison of its standards with the dimensions of existing buildings. No palace buildings are left, but in remote places some temples from the period can still be seen. During recent years some temples from the Song period have been found and measured. The temple Kaishan si in the county Xincheng, Hebei province, was built between 1004 and 1125. It was five bays long and should therefore have a unit of second or third grade. The cross section of the bracket arm decides the unit and the grade, but in Kaishan si the bracket arms have different cross-sections. The depths differ between 25.5 cm and 20 cm, and the widths between 16.5 and 15.5 cm. Kaishan si is closest to the third grade, which has a unit of 24 × 16 cm and a section of 1.6 cm.

The columns inside the building are placed independently of the columns in the wall to give room to the devotees (Fig. 20). At the bases the columns have a diameter of 53 cm, or approximately 33 sections, while the *Building Standards* prescribed 30 to 45 sections. The corner columns are 11 to 14 cm, or 3.25 to 4.48 inches taller than the columns at the center of the front side. This measure corresponds to the rule: "In a building five bays long the difference between the height of the columns is four inches." (*Building Standards/Yingzao Fashi*, Chapter 5, p. 48.) The slant of the columns toward the center of the building is 1 : 100 (Fig. 21). There are no levers in the bracket sets; the nose is the end of a beam, and there is only one inter-

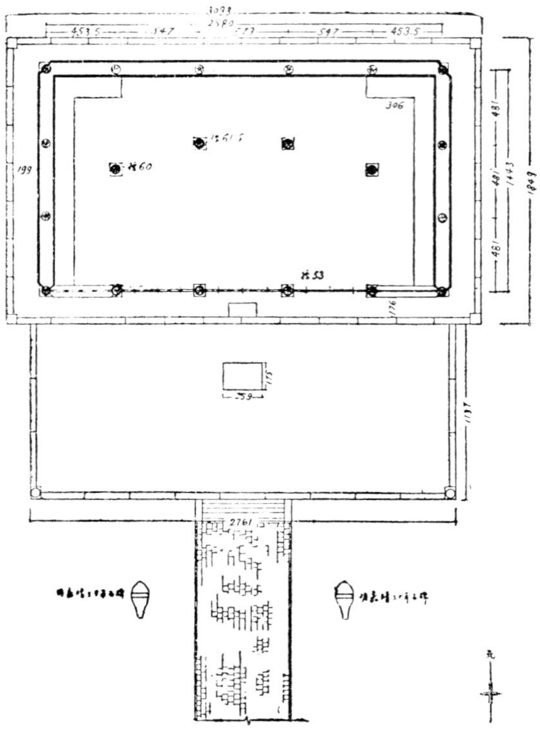

20. *Ground plan of Kaishan si.*

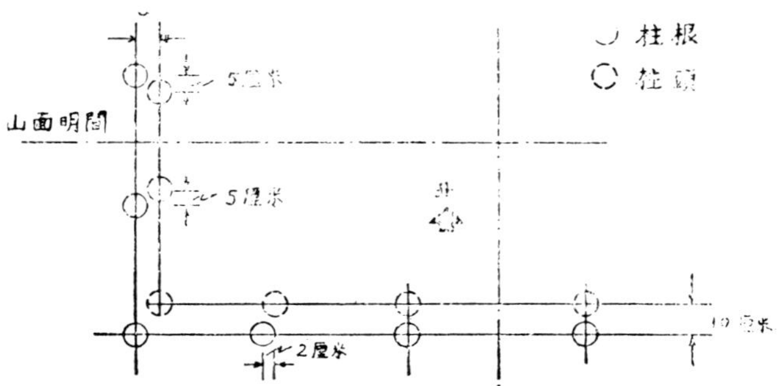

21. *Slant of columns in Kaishan si.*

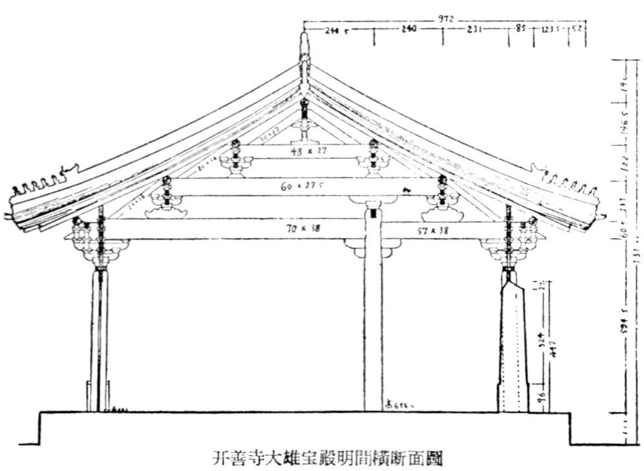

22. *Cross section from Kaishan si.*

columnar bracket set, also in the central span (Fig. 22).

Remaining buildings from the Song period are, to a great extent, built in conformity with the rules in the *Building Standards*, but no one follows them exactly. It is the dimensions of the bracket arms especially that differ from those of the codified standards. The standardization in units and sections was a living tradition, not a desk rule, and was modified by each group of craftsmen as they saw fit.

Later Work

A number of construction methods changed in the centuries after the Song dynasty. In later buildings the interior columns were placed in line with the columns in the wall in a regular grid. Both the increasing height of the columns toward the corners and their slant disappeared with the result that the roof curves became less vigorous. The oblique struts supporting the purlins disappeared, too. The frames of later periods were more rigid, with members joining each other at right angles.

The greatest changes occurred in the bracket sets carrying the eaves. The lever lost its shape and function and became a purely decorative element (Fig. 23). The eaves of later years were not as deep as they were before. In one of the earliest existing Chinese buildings, from 782, the ratio of the height of the bracket sets to the height of the column is 1 : 2. In the *Building Standards* the ratio is 1 : 2.5, but in buildings from the end of the Quing period it is only 1 : 6. The number of intercolumnar bracket sets was, according to the *Building Standards*, two in the central span and one in each of the other spans. The audience hall in the Imperial Palace in Baijing was restored in 1734. It has eight bracket sets in the central span and five in each of the others (Fig. 24). In early buildings the bracket sets created a most imposing collar around a building, but later they became merely decorative elements. There are different theories about the reasons for these changes. One is that the craftsmen gradually learned to make better joints and the bracket sets should thus have become superfluous. Another is that a gradual degeneration of method took place.

A modern Western engineer may find the traditional Chinese timber buildings unnecessarily complicated. Several Western art historians have speculated about the derivation of the Chinese curved roof. The most common theory is that it may have originated in the roof of a tent; there may, however, be more simple, technological answers.

Timber as a material has several advantages over steel and concrete. White cedar, the wood most common in Chinese buildings, has, by weight, a tensile strength about four times that of steel, and its resistance to compression is about six times that of concrete.

A building should resist vertical load from rain and snow, but also the horizontal forces of earthquake and typhoon. External vertical load is less than the great dead load of the tiles embedded on mortar in the roof. It is, according to the *Building Standards*, 400 kgs per square meter, measured horizontally in a big hall, and 280 kgs per square meter in a smaller building. In a Western building the weight of tiles is calculated as 100 kgs per square meter. The weight of the Chinese roof is clearly and easily transferred by the interior and exterior bracket sets to the columns and from these to the foundation. The Chinese frame with its short rafters is much more flexible than the Western frame with its rigid triangular truss. If the foundation of a Chinese building should subside under a column or two nothing would be forced to break.

A timber frame with many joints and extremely few nails is well suited to resist the oscillations of earthquake. The vibrations are reduced through the internal friction of the joints. Although the building will sway a great deal, it does not resonate with the earthquake, because it has a considerably longer period of vibration. Through deflection and friction in its joints, a wooden building absorbs the high frequency energy. Furthermore, the columns are not anchored in the foundation or in the stone bases. This means that the building can sway horizontally without breaking. The building is stabilized either by an open colonnade (Fig. 2) or by fixing an interior column (Fig. 22). Both methods insure a long period of vibration. The bracket sets and ties in the roof frame brake horizontal movement and reestablish the original balance.

The fragile walls are protected by deep eaves. A building with deep eaves and columns not anchored in the foundation is vulnerable to the surge of typhoons, but the heavy load of the roof counteracts the forces of a typhoon. This may have been the original purpose of the very heavy roof. The complicated bracket sets distribute the load of the eaves uniformly and that, too, makes them less vulnerable. The curve brakes the weight and provides the most efficient draining. The short rafters increase the flexibility of the frame, and are readily formed to a curved profile. The evidently aesthetic effect of the curved roof may have originated in a purely rational construction method. A structural analysis of columns and beams discloses the fact that the factors of safety of these members are similar to the factors of safety employed in modern construction. The durability of the traditional Chinese building technology is proved by the number of buildings that have survived typhoons and earthquakes since the ninth century.

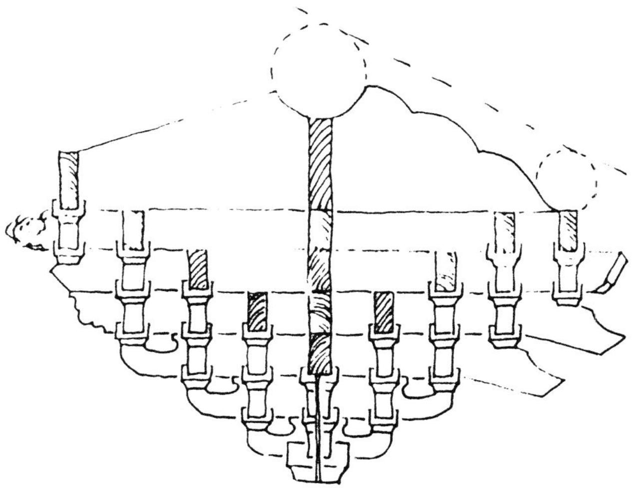

23. Bracket set from the 17th century.

Front of a building from the 10th Century

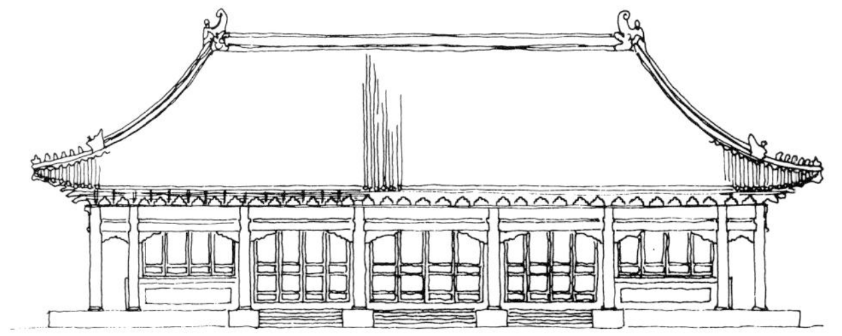

Front of a building from the 18th Century

24. A building from the 10th century and one from the 18th century.

BIBLIOGRAPHY

1. Li Mingzhong Yingzao fashi, 1925.

2. Liang Sicheng: Zhongguo jianzhu zhi liang bu "Wenfa keben." Zhongguo xueshe huikan, VII, 2. Sichuan 1945.

3. Liang Sicheng: Song Yingzao fashi tuzhu, n.d. Beijing.

4. Chen Mingda: Yingxian muta. Beijing 1980.

5. Chen Mingda: Yingzao fashi damu zuo yanjiu, Beijing 1981.

6. Pan Guxi: "Yingzao fashi" chutan, 1–2, Nanjing gongzue yuan zuebao, no. 4, 1980, and No. 2, 1981, Nanjing.

7. Guo Daiheng: Lun Zhongguo gudai mugou jianzhu di moshu zhi. Jianzhu shi lun wenji, Vol. 5, Beijing 1981.

Editor's note: An earlier article on this subject, "Chinese Building Standards in the 12th Century," by Else Glahn, was published in *Scientific American* (May 1981).

1. Interior of living room.

THE EXPRESSION OF MEANING AND THE NECESSITY FOR INTEGRATION

David C. S. Polk

THE CONTENTS OF CHARLES DARWIN'S NOTEBOOKS expressed the realization that man is *of* the earth, that our relationship to the world is not merely that of God-shaped passengers on a miraculous spaceship, but rather that man began, with all life, in primordial seas, crawled ashore with the amphibians, climbed and ran with the apes, and finally built Haghia Sophia, sang Bach in the cathedrals, and learned the secret of the stars. The evolution of life, from ancient seas to modern man—the man of recorded history—is a continuum millions upon millions of years long. We are separated from the process of natural selection only by time.

It is this fact, which we are not in the habit of contemplating, that lies at the core of the necessity for integration. Among the most fundamental human desires is the desire to be at home in the world, to feel ourselves to be in harmony with others and with the world at large, to feel the joy of living. In order to be at home in the world one must understand something of one's own nature and learn something of the nature of the world itself, the place of our origin. These understandings are the basis of meaning in life, and art is the intense expression of meaning, making meaning vivid. Architecture exists because of man's desire to express meaning in the shaping of his environment, i.e., to express himself, his innate humanness in the making of built spaces that can accommodate the events of his life with ease and grace.

Meaning is, by definition, common. It is available and intelligible to those who make the effort to reach it. The specific meanings of signs and symbols invented by the people of one culture may be obscure to those of another, but expressions of the common characteristics of human nature, such as a smile, a gesture, a touch, a song, a family story, a community legend, a neighborhood festival, etc., span the gap and render their meaning accessible (Figs. 2–5). Meaning in architecture lies in the truth of its reflection of human nature and in the integrity of its expression. The great achievement of genius throughout history is the clarity of the revelation of that nature, the revelation to man of himself. An understanding of human nature is the one constant that must guide and inform all architectural work.

Observation reveals the prevalence of integration in the natural world. The parts of an organism in nature are inseparable and are integrated into a complete living being that has characteristics and can be named. Parts are not interchangeable among organisms. Shape and substance are a direct physical expression of purpose. If I say that the image in Figure 6 is that of the leaf of a tulip tree, I am stating a simple fact with which no one will disagree. The general shape, color, and substance of the leaf is a perfectly integral expression of the fact that it is a miniature factory in which simple carbohydrates like sugar are made from small quantities of carbon dioxide present in the atmosphere. This process, known as photosynthesis, is the fundamental foodmaking process in nature, without which no plants or animals could exist. It requires the action of direct sunlight on chlorophyll, the green substance contained in the middle layer of cells in the cross-section of a leaf, which gives the plant world its characteristic color. The general shape of the leaf and its method of attachment to the tree give it access to sun and air. They belong to the phenomenon *leaf*, while its characteristic shape belongs to the specific organism known as the tulip tree.

2. *African Women.*

3. *A bridal pair at a café on the Marne which is part bistro, part garden. The two were here for the entire afternoon with a full wedding party which included uncles, aunts and small children of the family. (Photograph by Henri Cartier–Bresson at Joinville-le Pont, near Paris, 1938.)*

If I say that the room shown in Figure 7 represents the institution *museum* or that shown in Figure 8 is a library work room, it is again unlikely that my remark would provoke disagreement. Human purposes are very complex and the spaces made to house them must satisfy psychological as well as functional needs and desires. Yet a good room is also an organism, so appropriate for its purposes that one finds, on entering the space, that the potentiality of those purposes being fulfilled is invoked in one's mind. Space becomes place and, as Louis Kahn said, the space has earned its name.

The rooms of the Kimbell have indeed earned the name *Museum*. They tell the observer that they are good rooms and good places for looking at paintings and sculpture. Once the idea of the room was conceived, the act of creating the light of the room and the act of creating the room itself were one and the same. Space, material, light, and structure are all inseparably integrated, used in a way vastly different from the use of "devices" for admitting and controlling daylight. The Kimbell Museum integrates the character of the built space with the nature of the human activity to be housed. It integrates the creation of light with the creation of space, the nature of material with the structure of space and with methods of construction, and it integrates the parts of the organism with the whole.

In the Bibliothèque Nationale, Labrouste's great Salle de Travaille represents an achievement of similar quality in the making of a library. A large space is defined and modulated by nine delicate plaster domes supported on sixteen slender cast iron columns. A gentle light comes from nine occuli, one in the center of each dome. The space is then enclosed by a massive arched masonry wall containing books and narrow balconies giving access to those books on three levels within the wall. Great tables on the floor provide the working surfaces for hundreds of people. The black table surface is divided by strips of wood, set flush in the tabletop, that define a work space for each person along the length of the table. Lamps with alternating color shades provide individually controlled lighting. The single person and the collective

4. *"A peasant funeral. According to custom, the priest who officiates at the entire ceremony leaves the group halfway to the burial ground. Only the bearers and mourners accompany the body up the hillside." (Photograph by Henri Cartier-Bresson at Lucania, Southern Italy, 1952.)*

5. *Halloween Party—Germantown Friends School, Philadelphia.*

6. *Tulip tree leaf.*

7. *Galleries, Kimbell Museum, Ft. Worth, Texas by Louis I. Kahn.*

8. *Work Room, Bibliothèque Nationale, Paris by Henri Labrouste.*

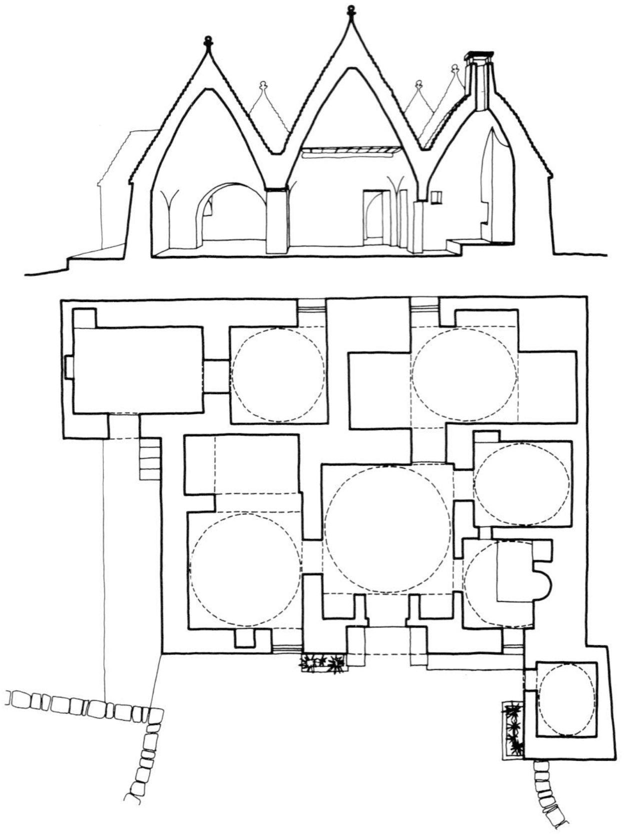

9. Plan and section–farmhouse near Selva di Fasano, the Murgia of the Trulli, Southeastern Italy.

10. Dooryard of farmhouse shown in figure 9 "where most of the household chores are done." (Caption quote and photo by Edward Allen.)

whole are marvelously defined and provided for in the same environment. The room is full of the presence of what Aldo van Eyck calls twin phenomena, without which there can be no human content in a space, no sense of room. The room is simultaneously delicate yet massive, big yet small, light yet dark. It is both high and low, broad and narrow, expansive and intimate. It is inclusive in character and is thus like a fragment of the world itself. One feels at home in it and can recognize it as a true expression of the institution *library*. In both these manmade places, as in the case of a natural organism, the character, shape, and making of the organism are thoroughly integrated.

Integration in architecture can be achieved on four levels. The first level is the integration of the shape and character of space with the specific and general nature of the human activities to be housed: the question of the discovery of what the character of a place ought to be; of what to do. As Kahn said "It is better to do the right thing badly than the wrong thing well." Integration on this level is required if a building is to have meaning. The right decisions here determine its basic habitability, whether or not its spaces will be truly appropriate, a graceful encouragement to the lives of its inhabitants, and whether or not it will be psychologically satisfying.

The second level is the integration of the shape and character of space with the nature of the materials used to construct it: how to *express* the character of the place. Integration on this level is required if a building is to be readable, i.e., its meaning made clear through the language of its construction. The language of architecture is the integrated structure of space. The unmortared stone trullo houses found in southeastern Italy show the power of the imagery that comes to a work of architecture through the integration of the shape of space with the inherent potentialities of the materials of which it is built (Figs. 9,10).

The third level is the integration of the parts of a design with the whole–of the spatial qualities of rooms with the details of construction employed in the making of those rooms, the grammar and style of a work. This level is to be distinguished from the previous one in that it refers to an individual architect's vision with regard to the specific nature of that work. It concerns his own personal means of expressing his spatial intentions while maintaining the respect for the inherent nature of the materials with which he builds demanded by the second level of integration. For example: Lutyens, in the Deanery Garden (Fig. 11), Kahn at Exeter (Fig. 12) and the builder of a caravanserai vault in Isfahan (Fig. 13) all express their individual predilections in design while exhibit-

11. Deanery Garden by Sir Edwin Lutyens.

12. Library, Phillips Exeter Academy by Louis I. Kahn.

ing profound respect for and understanding of the nature of brick itself, hand sized pieces of fired (or sun-dried) clay well suited to the making of foundations, walls, piers, arches, and vaults—a compression material made of earth itself. Each of these works has its own grammar, its own style. Each is an integral whole with which its parts are in harmony. They are very different in design, but all acknowledge the principle of integration. They are similar with respect to the second level of integration and different with respect to the third. Knowing correct grammar is as essential as knowing the language to the full expression of meaning.

The fourth level of integration is that of the organism and its relationship with its environment. A building must be *of* the city, not just in it, *of* the landscape, not just on it. Integration on the fourth level is required if the environment as a whole, natural and manmade, is to be a habitable place. It is a response to the human need to be at home in the world. One's daily experience is a continuum in which events are not meaningless, disconnected fragments. Experience is a state of mind and understanding created by the influence of events upon each other in the mind and by one's reaction to them. A building in the world is part of a spatial continuum and must act in a reciprocal way with its environment. The sequence of

13. Hajji Karim Khan caravanserai, vault over main timcah.

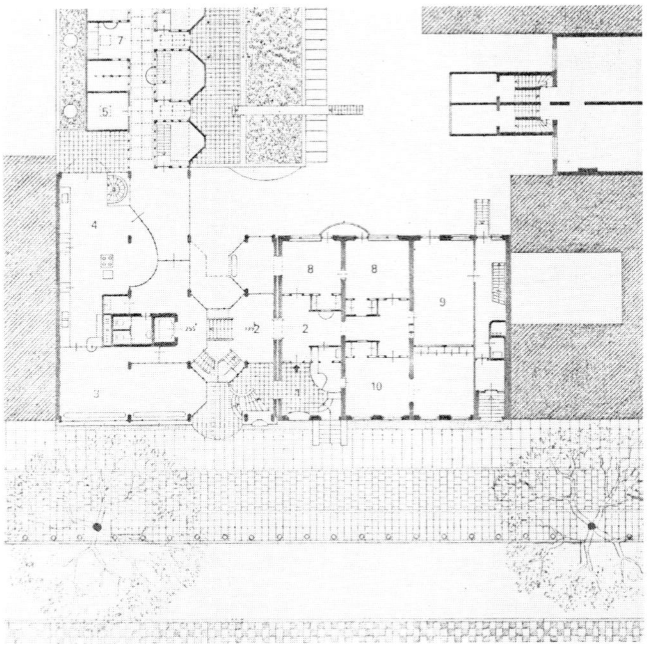

14. Plan, 1st floor, Hubertus House by Aldo van Eyck.

15. Sidewalk–new porches within the building–Hubertus House.

spaces from outside to inside must be a reflection of the continuity of one's experience in moving through them. Space must be defined by space as well as by material.

The manner in which one enters the Hubertus House, van Eyck's house for single parents and their children in Amsterdam, is eloquent testimony to the meaning of reciprocal interaction of a building and its environment, of building and city, and of the old and new parts that form a whole, singular entity (Figs. 14,15). At street level, both the old and new wings are hollowed out, forming a multilevel porch connecting old and new to each other and to the street, of which they become an integral part, offering welcome and shelter to passers-by and guests. Along the sidewalk, a seat is provided. A few steps up, in that portion of the porch that belongs to the old part of the building, a broad open windowsill offers another resting place. A new party wall is built adjacent to the existing one and new openings are constructed in this now especially deep wall, on axis with several of the important spaces in the new wing. From this new porch, a visual dialogue is established with the cafeteria, the main room overlooking the street on the mezzanine level in the new part of the building. The dialogue is created by locating the opening in the party wall on axis with the bay window at the east end of the cafeteria and on the center line of the structure of that room. This axis, parallel to the street, draws the two rooms (porch and cafeteria) together in the mind of the observer. A second axis, perpendicular to the first, includes the main interior hall in the new part of the building in the composition. Thus, the main entrance porch is a place where one feels simultaneously the presence of the street, the porches, the interior entrance hall, and the cafeteria. The sense of *interior* is enriched and extended to include both the street and the building not yet actually entered. The porches are, themselves, interiors, and one senses the interior of the cafeteria because of the transparency of its enclosure. Building and street are inseparable, and the street is given a new sense of presence by the building. Entry, both ways, is generously housed.

I have described integration as though it occurred on four different levels. The reality is, of course, that these four levels simply represent different aspects of one thing, which I have temporarily separated in order to clarify the nature of the idea. In the following pages, I will discuss these levels of integration with respect to the renovation of a house that we have recently completed in my office as they actually occur in building–as overlapping, interrelated phenomena. I will focus on the renovations concerning the entry and main living spaces.

The clients, after a long search for open land on which to build a new house, discovered a beautiful piece of rural property about forty miles west of Philadelphia that met all their expectations except that it already had a house on it in the prime location overlooking two ponds and a stream in front and grassy hills in the back. Privacy and future outlook were assured. The house, which had been built in the early 1950s, was organized in a way that suited the clients very well, and it contained approximately the right quantity of space (Fig. 16). Yet they disliked it intensely as a place and bought it, after considerable thought, in the belief that I could make something good of it. Somewhat to their surprise, the project was immediately fascinating to me, since it offered a classic opportunity to address an issue I find of continuing interest, that of the distinction between organization and architecture, between the diagram of a plan (a scheme) and a true plan that represents an inseparably related family of rooms.

The entry drive led to a large motor court, well-defined by hedges on three sides. The fourth side was poorly defined by the house itself, an uncoordinated, inarticulate collection of elements that had no sense of being the fourth side of an important room of entry (Fig. 18). There was no resolution of the question of scale posed by the garage door. The position of the front door was weak and incidental, and the door itself offered no sense of welcome and no accommodation to the function of an entry, such as a protection from weather, a place to rest packages, to collect oneself, etc.

The interior entrance corridor was long, narrow, and dark, with neither a sense of place in itself nor a sense of leading to a place in either direction (Fig. 20). Entry from the garage, very important in a country house, took place through the closet, up dangerous, winding, dark steps. There was a lean-to greenhouse, in poor condition, adjacent to the entrance corridor, but unrelated to it. It played no role in the experience of entry.

The interior living spaces, living room, dining room, and kitchen had been designed as a group, with dining and kitchen raised two feet above the living room, allowing them to be open to it while maintaining a degree of privacy, but the dining space itself was too small for our clients' needs, and the kitchen required new equipment. The living room was good-sized but lacked scale, and its light, while adequate in quantity, was one-directional and had no sense of belonging to the making of the room. The exterior living spaces were not well related to their interior counterparts, and the porch outside the dining and kitchen areas was covered with green fiberglass, giving everything and everyone below a greenish cast.

Thus, the first level of integration, the character of space with the nature of human activity, was minimally fulfilled by the spaces of the original house. None of them had the quality of being rooms, i.e. they lacked a sense of completeness, of being composed of integral parts, of being real interiors rather than merely enclosures and of being *evocative* of ap-

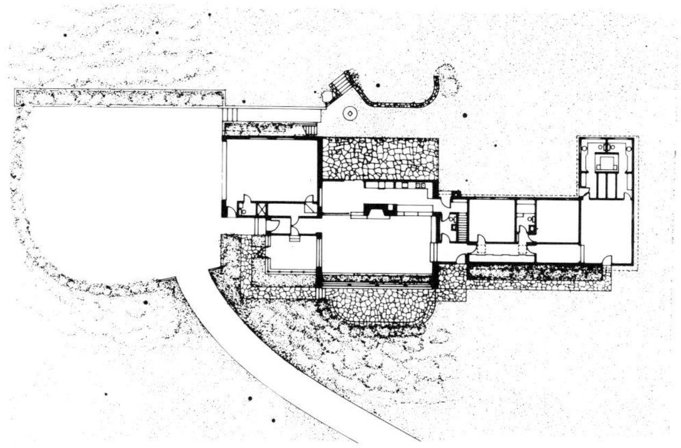

16. Plan of original house, as found.

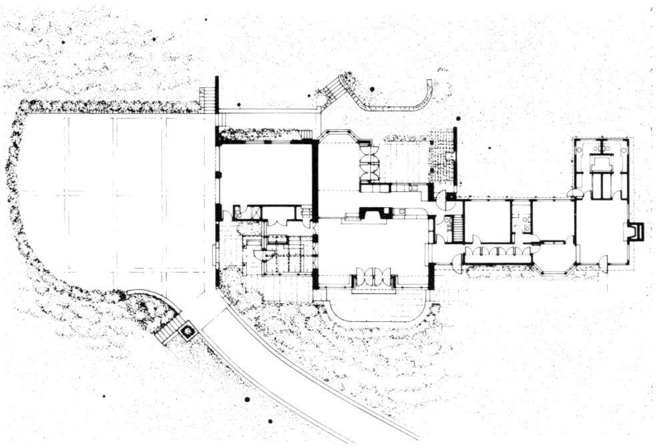

17. Plan of renovated house. Additions and alterations are shown in black.

18. Entrance court elevation—original house.

19. Entrance court elevation—renovated house.

20. Entrance "hall"–original house.

21. Living room–original house.

propriate activities.

The second level of integration (of the shape and character of space with the nature of materials) was not well served by the manner in which the building had been made. The principal materials were stone and wood. Openings were made in stone walls without regard for the nature of the material, and nearly white mortar joints outlined each stone at the expense of the expression of the wall as a whole. Wood siding was carried straight to the ground, where it had naturally rotted. Wood roofs overhung the walls erratically by different dimensions, none of which made sense either as protection for the windows or as the extension of interior space into the landscape. Ceiling joists were exposed in some places and plastered over in others in a similarly haphazard manner. The slope of the roof was just enough to be disquietingly not flat, but not enough to be a pitched roof. The living room fireplace did not draw, since the chimney was not tall enough to provide an adequate column of hot air. However, the old sliding window sash, made with delicate horizontal muntins, was very nice, and the windows were generally well-located.

With respect to the third level of integration, there was no sense of the relationship of part to whole, of the role of the detail in relation to architectural intention, because there were no details, and possibly no intentions. Windows were simply cut into walls without concern for the fact that sill, jamb, and head constitute three different conditions. Windows had no sense of specific position with regard to the quality of light entering a space or with the expression of an idea concerning the relationship of interior and exterior spaces. Overall, there was no expression of the nature of materials through the means of construction.

Finally, concerning the fourth level of integration, although the building had been very well sited with regard to its position on the land and general orientation, no explicit relationships had been established between interior and exterior spaces by means of the manner of construction of those spaces and of the ways in which they were connected. In addition, although the building was one story high, no advantage had been taken of that fact. The roofs played no role in helping to clarify and express the different qualities of the various spaces within and, for the most part, gave no indication to the inhabitants that the ceilings were, in fact, roofs with sky above.

In order to make a building of this building, to introduce it to the world of architecture as it were, I proposed that we make many changes. Beginning at the entrance, we gave the house a new front wall, one that would remake the fourth side of the motor court and, at the same time, tie the house to its land by extending beyond it on both sides to frame new gardens (Fig. 19).

This wall is of concrete block faced with stucco. Four openings are made in it, each one constructed of stone in a manner related to its purpose. The stone used is the same as that in the existing house. The contrast between the new stone-built openings and the stucco surface in which they stand emphasizes their nature as constructed elements in their own right instead of as merely areas where the wall has been removed. The main entrance opening, at the right hand end of the wall, is framed by a stone arch spanning nine feet at its spring points with a height of seven feet at the center. The lower half of the opening, which is generally T-shaped, is four and a half feet wide, thus making a ledge at hand rail height on either side as one passes through the wall. The ledge provides something to grasp, if needed, or a place where the owner might set flower pots out in spring. Restricting walking space to the middle of the opening allows the springing of the arch to start below head height so that one walks within the arch rather than under it, making the experience of it more immediate and different from walking under a lintel or a beam. The opening is both wide and narrow and has several heights. It has scale and a particularity appropriate to the principal entrance of a house. Its construction is expressive of the nature of the material of which it is built.

The arch is axially aligned with the small new front porch just beyond it and with the new glass-roofed interior entrance garden which is the old greenhouse reshaped and rebuilt. The axis is centered on the existing opening between this new interior entrance garden and the living room beyond. That opening is thus both symmetrically axial from the entrance side and incidentally asymmetrical from the living room side, since the center line of the living room is shifted slightly off the entrance axis. One is always aware of this shift in axis from entrance (exterior) to living space (interior). Exterior and interior are thereby brought together in reciprocal interaction, both present simultaneously in the consciousness of the inhabitant. The shifting of the living room off the entrance axis reflects the fact that it is a destination to which the entrance sequence leads, rather than just another part of the sequence itself. This sense of destination is further reinforced since one must turn off the axis in order to enter the room.

The portion of the new front wall in which the entrance arch is centered is articulated from the garage by a notch (Fig. 23). The sill of this notch is the same height as the ledges within the arch. That height continues beyond the house on the other side as the wall is extended to enclose a portion of the garden, making a horizontal line that helps to unify the composition of the elevation while further uniting building and landscape. The space between the new wall and the living room is "inside" the house and, although a portion of that space is outside, it is all an "interior." One enters the house before reaching the front door. A sense of welcome is conveyed by a building when it freely offers accessibility to "interior" spaces, spaces not behind closed doors, places that seem to anticipate and accommodate the needs and desires of those who are arriving.

The existing garage door was replaced by a pair of smaller openings spanned by cast-in-place concrete masonry lintels that are fitted with fully glazed doors. These openings are more in scale with those of the house than the original large one and give the garage the feeling of being a room in the house. A new and more reasonable entrance leading directly to the house from the garage is provided. It is lit by a skylight and has a ledge beside the door for resting packages.

At the right-hand end of the wall an opening with no spanning element leads up a few steps to the gardens and hills in back of the house, a simple gateway in a garden wall. Again the wall extends beyond the house into the landscape. It is both the wall of the house and a garden wall.

The experience of entering the remodeled house is accommodated by a series of well-defined spaces through which the outside world and the building act integrally with each other. Arrival at the entrance court is signified by the beginning of the garden wall and a massive stone flower pot opposite it that ends the hedge. The court itself is given a floor by means of strips of cobblestone paving set in a pattern in the asphalt. A wider strip marks the entrances to the court and to the house. This new floor, along with its new wall, gives the entrance court a much greater sense of place than it originally possessed. The new floor has an elevation, a horizontal façade to the sky. After passing through the arch, one enters a garden that continues around the southwest side of the house to the living room terrace. A wood trellis, to be vine-covered, is attached to the inside surface of the wall—a wall garden inside, a garden wall outside—providing a large surface of greenery in a very small space. The new front porch, which shelters the front door and a wood bench beside it, also forms a portal through which one must pass in order to reach the terrace beyond (Fig. 24). The feeling that the terrace is a special place is enhanced by one's entering it through the beginning of the main entry sequence.

The ceiling of the entrance porch is detailed as though it were an interior space with plaster set flush within wood edges, while the entrance conservatory (the first indoor space) is glass so as to give as much a feeling of openness as possible, as though one were outside again in a court (Fig. 25). The floor of this room is paved with slate, as are the exterior terraces. One then moves up three steps to a landing (the old entrance corridor opened up) that overlooks the conservatory like another porch, only this time one that is inside. An upholstered oak bench forms the "porch rail" and provides a seat for taking off boots, etc., by the closet. From this space one enters the living spaces of the house through an existing opening in the original fieldstone end wall of the living room.

22

23

24

25

26

22–26. These photos show the sequence of spaces through which one enters the renovated house.

27. *Section, original house.*

28. *Section, renovated house.*

This sequence of spaces is intended to make the building and its landscape inseparable, each extending into the realm of the other, while providing the appropriate places of entrance. The quality of place is the reality of architecture. Symbolism can enrich meaning, but symbol without place is empty irony.

The living space consists of the living room, kitchen, and two terraces. The original ceiling of the living room was removed and replaced with an oval vaulted ceiling made of 1×4 cedar boards (Fig. 27, 28). The vault opens the ends of the living room to light without increasing the height of the side walls. The shape is volume-containing, and it gives that volume to the light of the room. The shape of the space and the quality of the light become integral with the making of the room.

The original 34-foot-long plant box that bounded the entire southwest side of the living room and separated it from the exterior terrace beyond was removed, along with the glass wall above it, and was replaced by an arrangement of windows and doors designed to connect the living room with the terrace and also to provide low-ceilinged alcoves deep enough to sit in along the edge of the living room (Fig. 29,30). The ceiling of these little "rooms within the walls" is a horizontal plane, which casts a shadow modifying the daylight at the edge of the room. The height of this plane is also the height of the new doors and operating windows, and was chosen to align with the height of the existing openings in the stone end wall. Above that line, a band of fixed glass encloses the room to the soffit of the spanning elements that support the vaulted ceiling.

The glass doors in the center section are set back at the inner face of this four-and-a-half-foot-deep wall so as to protect them from the weather and give them a place within the wall when they are fastened in the open position. This space forms a narrow and relatively tall porch connecting the living room with the terrace. Its height contrasts with the very low alcoves on either side of it in the room and relates more to the sense of scale of the exterior.

Tall casement windows are placed in the corners of the alcoves, leaving the center glazing fixed. The window sills are set about 12 inches above the floor, increasing the sense of intimacy in the alcoves.

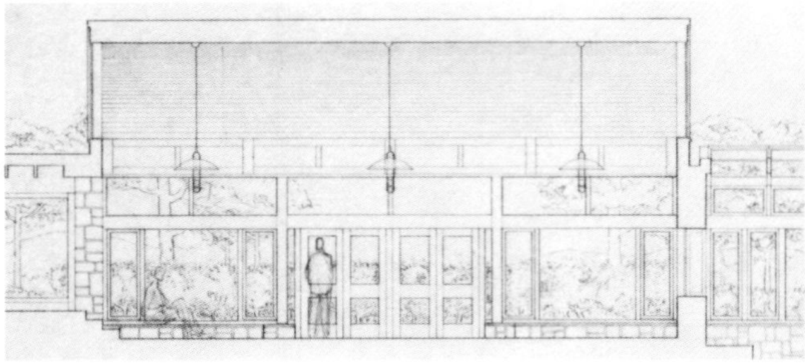

29. *Interior elevation of southwest wall of living room—renovated house.*

30. *Living room looking southwest.*

31. *Relationship between living and dining rooms.*

The entire deep wall of doors and windows is thus composed of three zones in height—the zone of doors and operating window sash, the zone of fixed glass, and the zone of the box beams that support the joists framing the vault-shaped ceiling. The vertical subdivisions in each zone vary according to its nature. The doors and alcoves divide the length of the room into unequal but symmetrical thirds, with doors and windows naturally having different rhythms. Immediately above, in the zone of fixed glass, those thirds are each divided in half by a single mullion. Above the fixed glass the boxbeam is sheathed in white oak style and rail paneling, which is divided equally along the entire length of the wall, first into four major panels that are, in turn, divided in half by smaller vertical styles. All three rhythms coincide at the center and ends of the room, but are otherwise contrapuntal, thereby emphasizing the different character of each zone and the fact that the wall of the room has many heights within its overall height. The volume of the vault itself adds still another dimension to the height of the room. The room, then, is simultaneously high and low, and it is a large room that contains very small spaces along its edges.

The low ceiling covering the original dining area and kitchen was rebuilt as a flat plaster plane trimmed with wood edges. The dining room was extended by removing the existing porch, and a new bay window was added that projects beyond the corner of the garage. This new portion of the dining room was covered with another smaller oval vault set at right angles to the living room so as to bring sunlight into the dining room over the flat-roofed section of the original dining area. The low, flat space between the two vaults is a kind of alcove to the new dining room that overlooks the living room, belonging to each but in different ways (Fig. 31).

The dining room vault is smaller, lower, and more intimate than that of the living room because, while they spring from the same point, there is a two-foot difference in floor level between the two rooms. A bank of glazed doors set in an alcove connects the new dining room to an open terrace beside it. This terrace is defined opposite the dining room by a new wall and pergola, which will be covered with wisteria. There is a small opening in the wall, framed in stone with a stone table made as an extension of its sill, a permanent part of the building. The pergola forms a little green-roofed room in the landscape at the edge of the dining terrace, the right size for two people. The juxtaposition of this space with the terrace both defines the limit of the terrace and adds to its quality

32. Dining terrace and pergola.

as a place.

The relationship of the kitchen to the rest of the living space was excellent for the clients as we found it. The relationship between the kitchen, and living and dining spaces is such that a person working in the kitchen overlooks, and has access to, the living room while from the living room the kitchen is not seen as a kitchen since no counter or equipment are visible (Fig. 33). Visual privacy is also maintained between the kitchen and the dining room.

The living space as a whole is thus composed of several interdependent yet clearly individually defined rooms and adjacent alcoves both indoors and out. Each space is defined by means of a differentiated overhead enclosure and by the horizontal juxtaposition of those spaces while maintaining openness from one space to the next. Space is enclosed by space; where one *is* is defined in relationship to where one *is not*. The space occupied is always seen as being part of a larger whole. It is the nature of living space in a house that it be seen in relation to the world beyond: one's own realm within the larger realm we also inhabit. It is an expression of the sense of house as a garden before it is a house, an enclosed and harboring place in the world, yet seen as part of the world.

The materials and methods of construction used play a major role in determining the character of the spaces in the rebuilt house. The scope of structure in most houses is limited but construction and detailing are critical to the expression of spatial ideas. A detail becomes meaningful when it is an integral part of both the spatial character of the building and the order of its construction and thus aids in the perception and understanding of these two facets of building.

The stone walls and the openings within them were poorly made in the original house. The mortar joints were too big and were of a color much lighter than the stone, so individual stones seemed to float in a field of mortar rather than make a wall. A wall is an element of architecture, not a collection of stones. The openings were made with concealed steel angles carrying the stonework. Thus, neither stone nor steel is allowed to express its inherent nature. The stone, chosen because of personal preference for its color and texture, is made to seem weak and pictorial. Stone is the material of which the earth itself is made and is worthy of respect. Similarly, the potentialities of steel for lightness and extraordinary tensile strength are not realized, and the steel members are reduced to the role of reinforcement.

The joints were raked and repointed with a mortar about the same color as the stone. The owners did

33. Kitchen looking toward living room.

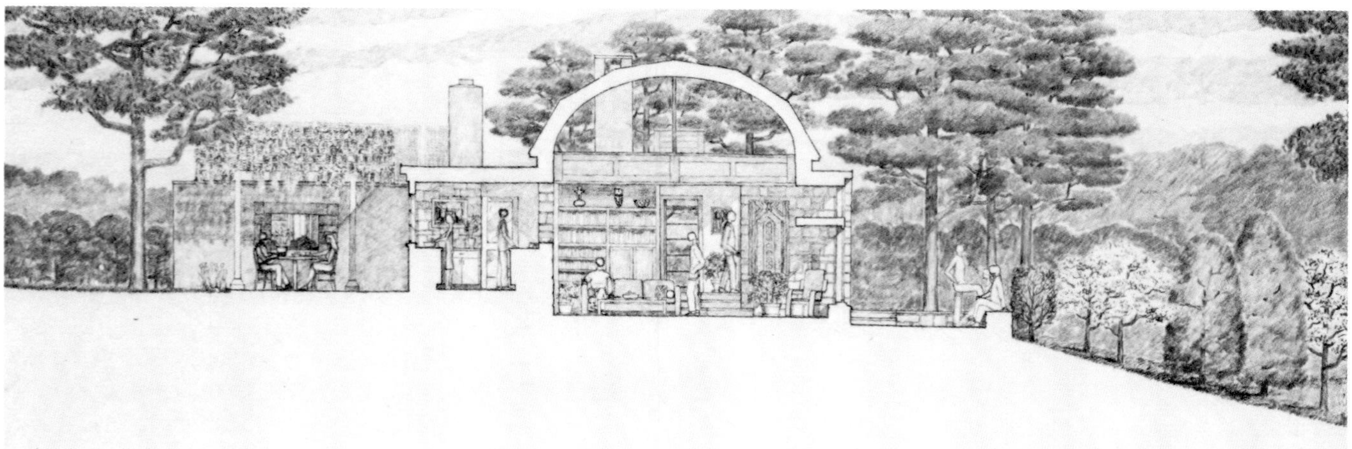

34. Section study through dining terrace, kitchen, living room and living room terrace.

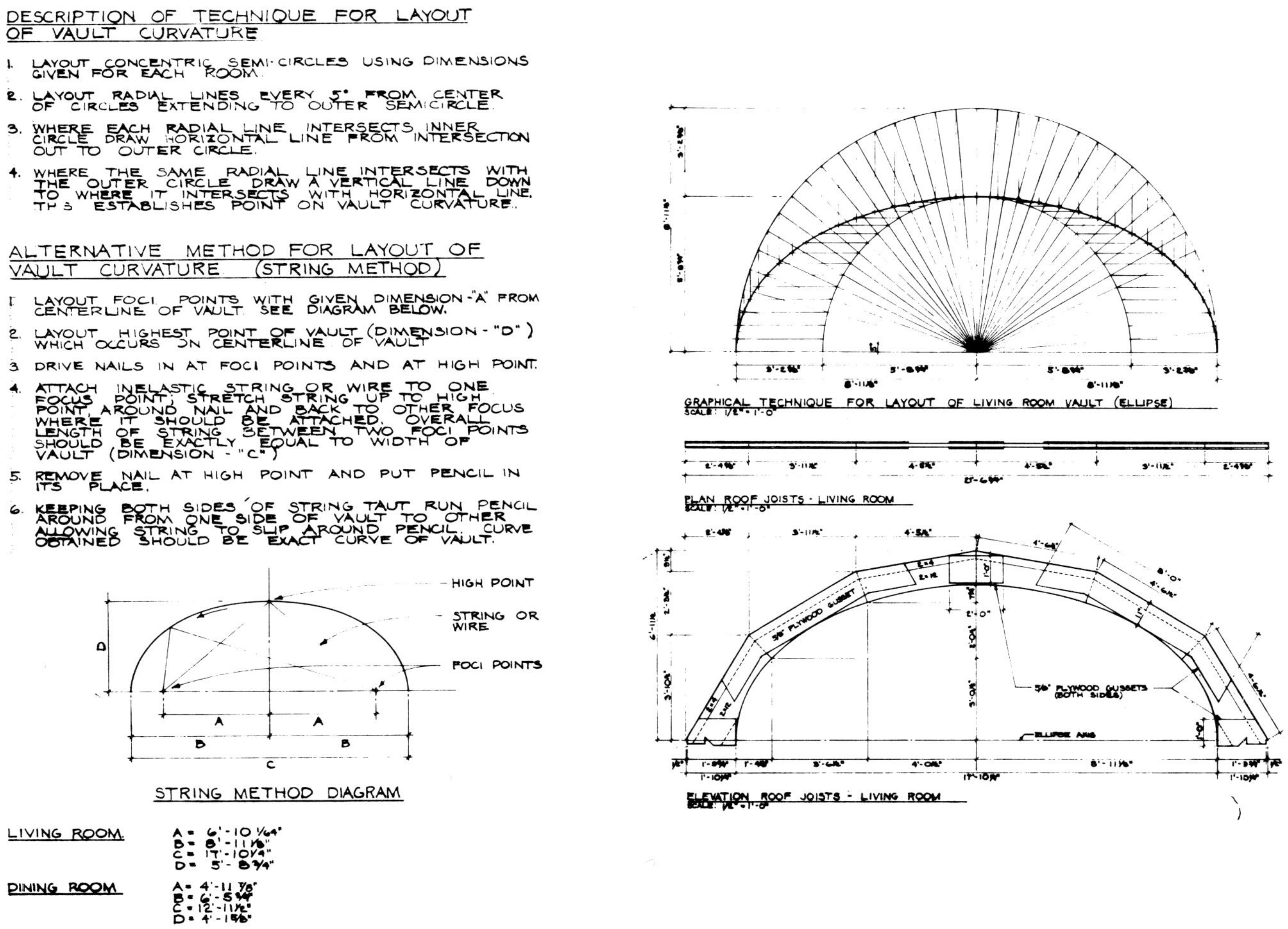

35. Working drawing—construction of roof joists.

not wish to rebuild the openings, so in order to give them the sense of being finished off in some fashion, we lined them with white oak. To close off the glass-roofed entrance room from the living room on cold winter nights and hot summer days, doors were made that fold into the lining of one opening and cabinets with shutters on top, which fold into the lining of the other. Furniture, in this case, becomes part of the architecture of the building.

The major piece of construction required for this transformation was the making of the living room and dining room roofs. These vault-shaped roofs are constructed in a manner somewhat similar to that used in making the hull of a wooden boat. The roof joists are made of shaped pieces of 2×12 material sandwiched between two layers of 5/8″ waterproof plywood gusset plates. The top surface is divided into six equal sloping segments that form a faceted gambrel roof suitable for fiberglass shingles. The curved outer surface, which I originally proposed, would have been difficult to sheath in plywood and would have required a metal roof, very expensive if one wants it to be maintenance-free. A fiberglass shingle roof would last thirty years at one-tenth the cost and be more in keeping with the rather ordinary character of the exterior of the house. The six-part slope of the exterior closely follows the shape of the oval ceiling below and is expressive of the fact that joists such as these are not continuous and must be made up of pieces. The whole composite joist acts as a structural unit with the 2× material in the middle providing the nailing surface for exterior sheathing and interior tongue and groove cedar boards. The contractor, who did a superb job of building, was given precise instructions for laying out and building the joists (Fig. 35). Dimensions had to be held with great precision, since the finish wood that covers the beams was required to be flush with existing surfaces.

The depth of the wall of the glass doors and windows is used structurally. The roof of this deep wall is constructed so as to act as a horizontal beam resisting the thrust generated by the joists under load. It acts integrally with its vertical leg, the supporting beam.

Window details are designed to intensify perception of the relationship between indoors and outdoors. Fixed glass is set flush with interior wood surfaces and held at the outer face by one 3/4″ trim piece, i.e., held at the outermost possible limit of the interior

36. Roof construction.

37. Entry under construction.

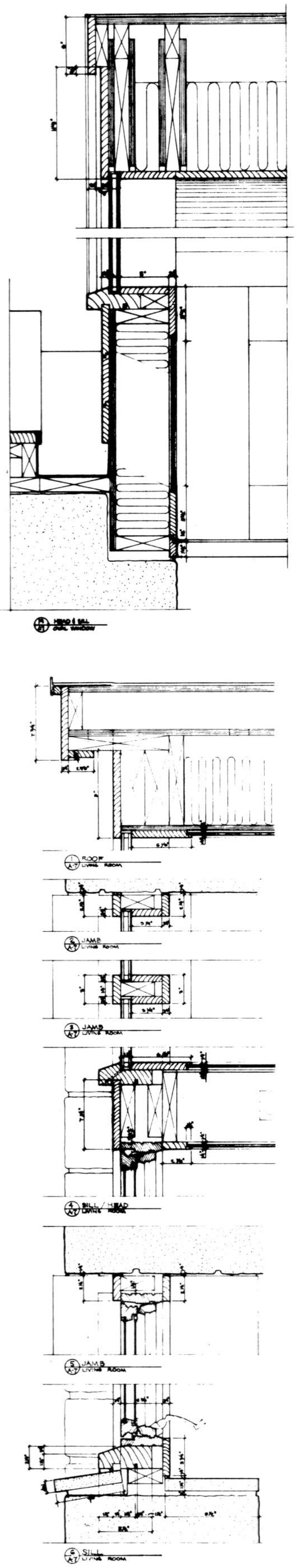

38. Oval window details–living room.

space (Fig. 38). This flushness at the outer limit combined with mullions designed like small free-standing posts with none of the usual reveals that mark the holding of glass, gives the presence of the glass an ambiguous quality. At first glance the space may appear to be open and unglazed, yet the position of the glass at the very edge of the building reinforces that edge and restates the fact that one is indoors. The effect is particularly clear at the ends of the vault-shaped living room ceiling, where the feeling of openness, of the nonexistence of glass, is clear. No shadow exists at the joint between glass and ceiling, thus reducing the contrast between the intensity (glare) of the light of the sky and that reflected from the surface of the vault. The shape of the vault is completely clear against the sky, and it sharply defines the edge of the room. The effect is of simultaneous clarity and ambiguity, of a dialogue between outdoors and indoors. It is another way of bringing the indoors and the outdoors, the building and the world, into communication with each other in a way that will be felt by those who occupy the space.

Built-in wood furniture, cabinets and bookcases at both ends of the living room, and rails and benches, are made of solid wood stile and rail construction framing plywood panels. These white oak veneered panels are edged with solid wood strips set flush within their frames. A 1/8″ joint allows the panels to move as they expand and contract. All finish surfaces of this furniture are flush and smooth to the touch.

The paneling at the base of the vault sheathes the beam and is then extended around the room. It forms the inner edge of the flat roofs, which are beyond both ends of the room. This paneling is set back 1/4″ behind the face of its stiles and rails, creating a thin crisp shadow line around each panel. Greater differentiation is needed between panels and surrounds the further away the panel is from the observer. The panels above slide behind their styles and rails rather than into them. White oak was used for the paneling and cedar for the ceiling so as to express the different conditions behind them. The oak, being much denser than the cedar, takes the finish in a different way and thus reflects light more.

The new interior floors in the living spaces are made of Mexican Satillo tile, a beautiful and inexpensive twelve-inch-square tile that is full of pink and yellow, colors that are also prevalent in the cedar boards of the ceiling, so that, although the ceiling and floor are different materials, they speak to each other in color.

In the remaking of this house, I have tried to express meaning in the making of space. I have tried to make rooms of what were merely spaces, to make a house of what was before little more than shelter. Meaning is accessible to anyone who is willing to make the effort required to try to understand the actual nature of things–human nature and natural phenomena. The ramifications of these basic components of life are endless and varied, simple and very complex, and the attainment of a degree of understanding is a life-long undertaking. Meaning always eludes those who oversimplify, those who over-complicate and those who seek appearances, stylistic and otherwise, rather than appearance, the true expression of a nature.

Integration in architecture is a necessary means to the end of the expression of meaning. Integration is a kind of description of the way the world has been built by natural process and natural selection and of man's position in the world. If one's work is to be intelligible to others, if it is to have meaning, it must respond to the necessity for integration.

Acknowledgments

Invaluable assistance in the work was given by John David Rulon, who also did the working drawings, Gautam Bhatia and Faith Baum all working in my office, and by Chuck Bloszies of Keast and Hood, structural engineers. The work of George K. Page, the contractor, and his subcontractors was excellent, particularly the onsite carpentry of Bob Miller, the foreman on the job. The clients' sensitivity and desire for quality were a great encouragement to us.

1. The Colonnade of the Louvre, by Claude Perrault, under construction in 1674, from an engraving by Sebastian LeClerc.

THE MECHANIZATION OF ARCHITECTURE AND THE BIRTH OF FUNCTIONALISM

Alexander Tzonis and Liane Lefaivre

IN THE EARLY YEARS of the postwar period, modern architecture was accepted all over the western world as the only means of implementing the vast reconstruction programs of the times. By the end of the 1970s the situation had changed. Modern architecture was universally declared obsolete, dismissed as a grave mistake, and blamed for practically everything that was wrong with buildings, cities, and landscapes. In addition, postmodernism, a complex of tendencies that seemed to renounce practically everything modern architecture embraced, emerged as the desirable alternative.

This reversal of opinion presents an interesting feature. In victorious moments, modern architecture was *praised* for being "machinelike," for being functional, efficient, rational; at the lowest ebb of its popularity it was *damned* for being machinelike. Now the label has come to mean ugly, insensitive, oppressive. "Machineness" seems to be at the heart of the matter. It has given rise to many arguments. But it has also tended to confuse. By focusing on this single point, the debate has overlooked other important issues.

The problem is, first, that the expression "machinelike" is used loosely and ambiguously and, second, that it applies to the fabric of the building only, ignoring the role of human relations and of social change in the emergence, evolution, and current crisis of modern architecture.

To think more clearly about the architecture of the years to come we need a more rigorous and broad understanding of the past. The best way to achieve this is to investigate the conceptual and institutional development of modern architecture—in other words, to turn to history.

The following paper on the relation between machine and building, on what we will call the "mechanization" of architecture, is written in this perspective. We feel that this development is central to the understanding of modern architecture.

Mechanization of Design

Mechanization of design can mean several things. It can be taken as the introduction of the machine onto the construction site, into the materials factory, and within the functioning building itself.[1] Understood in this sense, mechanization is the descendant of the inventions of the industrial revolution. Among its even earlier predecessors are Agostino Gerli's ventilation schemes for large indoor spaces,[2] Claude Perrault's hoisting apparatus for heavy construction elements,[3] Agricola's mining machines,[4] Domenico Fontana's devices for moving the Vatican Obelisk,[5] Leon Battista Alberti's suspension "engines" made up of "pins, levers, and pullies"[6] and, finally, in antique projects such as Hero of Alexandria's steam-powered temple doors.[7] (Figs. 1–6.)

Mechanization can also bring to mind the designers and the movements—Tatlin, Sant'Elia, Charot; Functionalism, the International style—that imposed the machine and the machine-made look as the chief aesthetic prototypes of twentieth century design.

The connection between the machine and ideology is much more rare, and it usually evokes the vision of the building working as a machine. This idea is contrasted to older, more "humane" approaches to architecture that viewed a building as a living body, an animate object. The famous dictum of Le Corbusier that a house is a machine to live in (Fig. 7)[8] is often juxtaposed to Renaissance plans of buildings on which

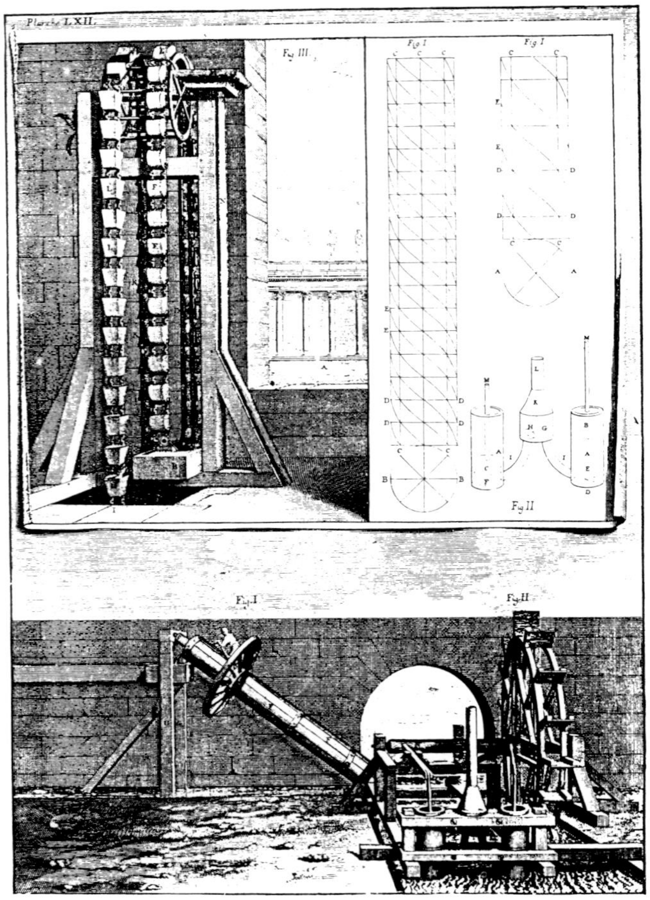

3. *Machine for raising ore from mines, activated by a water mill. Illustration from Agricola's* De Re Metallica, *Book* VI *(Basel: 1556).*

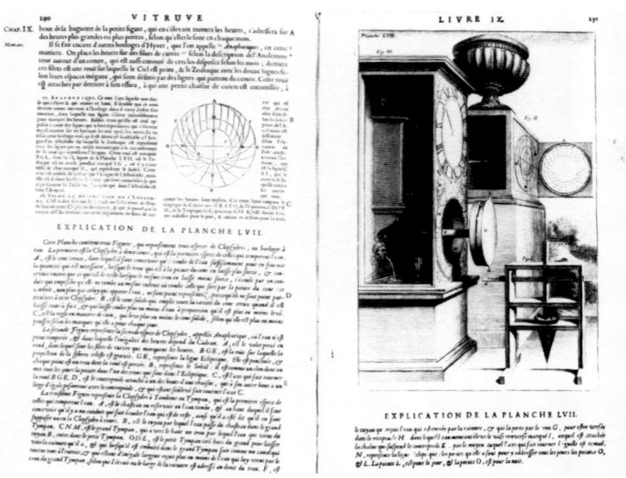

2. *Claude Perrault's illustrations of machines from his edition of Vitruvius's* Les dix livres d'architecture *(Paris: 1673).*

4. *An illustration from Domenica Fontana's famous book,* Della transportatione dell' Obelisco Vaticano, *(Rome: 1590), recording the 1586 transportation of a 681,222 pound obelisk. The operation was unique for its size and its complexity, for the machinery it involved, and for the number of people it employed in a coordinated manner. It required an organization equivalent to that of an army in battle. More than half a century later, Galileo still had the memory of the event, remarking in his reflections on scale in his* Due Nuove Scienze *(Leiden: 1638) that "a small obelisk or column or other solid figure can certainly be laid down or set up without the slightest provocation and that purely on the account of their own weight."*

5. *Leone Battista Alberti's suspension "engines" from his* Dieci Libri Dell' Architettura *(Florence: 1485), Book VI, Chapter VII.*

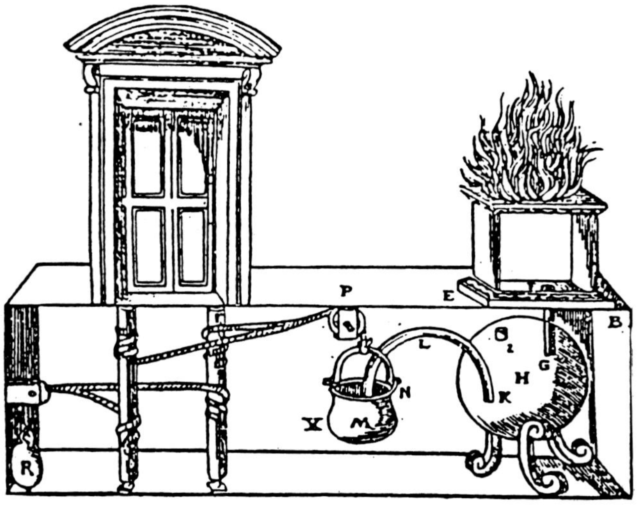

6. *Hero of Alexandria's mechanism for the automatic opening and closing of a temple door. From* Histoire des Sciences, *P. Brunet and A. Miele (Paris: 1935).*

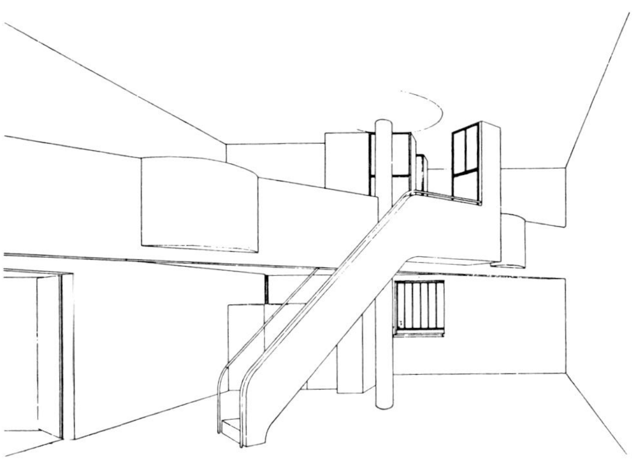

7. *Interior of a single dwelling in a mass housing project for artisans from Le Corbusier,* Vers une Architecture *(Paris: 1923).*

the human body has been drawn in order to determine the subdivisions and shapes of the plan.

This is a popular, impressionistic picture of the course of the development of architecture, and there is some truth to it. But it overlooks the fact that mechanization of architecture meant a transformation in the architect's thinking, in his perception of purpose, in his sense of the world, in his relation to the rest of society, in his methodology of design, and in the social use of his design products. It also harbors the false impression that mechanization was primarily a matter of technological innovation, industrialization, and the abundance of machines, especially around the second part of the nineteenth century. In fact, it was a fundamentally social phenomenon, and it occurred much earlier, towards the end of the sixteenth century and the beginning of the seventeenth.

Although the introduction of the machine into architectural technology, style, and thinking are related, its introduction into thinking is the most important of the three. Before the machine could be massively adapted in construction or become a standard of beauty, before in fact aesthetics could be perceived as distinct from a scientific approach, the thinking of the designer had to undergo a fundamental change, and the social use of architecture had to be altered. Both types of transformation we see as the essence of mechanization. Functionalism, which we discuss in the second part of this paper, arose at a later stage, to a large degree as an attempted synthesis between aesthetics and the scientific approach.

Methodological Note

Before discussing this change we would like to clarify the notion of conceptual systems in architecture.

The thinking of the designer takes place on three levels. The upper level consists of *norms.* These are prescriptive statements that the lower level prescriptive statements or *directives* resort to for justification or derive from as conclusions. Directives may be expressed verbally or in the form of drawings that are meant to direct design actions. Norms and directives are linked by *facts.* These make up the middle level. They are factual statements that specify that when the design directive is carried out, the norm is satisfied. Facts themselves need justification. This is supplied by the *backing.* It gives reason for accepting the facts as true. Finally, the backing needs support. This is provided by the *base,* which tells us why the backing is admissible.

What we have sketched here is the bare minimum of the conceptual framework of design, its kernel. Theoretically, all programmatic thinking takes place in the argumentation that links up these components.[9]

This conceptual framework replicates only schematically the chainlike character of design thinking, the elaborate network of argumentation through which architects arrive at their designs or through which they defend them. But the components of this kernel capture the fundamental elements of architectural thinking. This schema therefore suits our purpose, which is to describe the most significant changes that occurred in the conceptual system of architecture with the advent of mechanization.

Efficiency

Prior to the mechanization of design thinking, the norm of efficiency was absent. The norm of purity governed in its place. It prescribed the mixing of elements that enter into the making of an object—colors, materials, shapes, proportions. This norm dominated the thinking of designers in archaic or tribal cultures, as revealed in numerous documentations by anthropologists or in texts such as the Bible and the Indian and Chinese cosmologies. In medieval and Renaissance literature, this norm took the names of ratio, concordia, consonantia, and harmony as it moved toward eclipse and disappearance from the way of thinking.

Efficiency is one of the norms that took the place of the norm of purity in the *modern* [10] conceptual system of design as that norm of purity began to lose ground. Efficiency was defined as the search in the construction or operation of buildings for maximum benefit at minimum costs.

Optimization

Optimization, the satisfaction of two competing requirements, is an obvious component of design thinking to associate with mechanization. However, it makes its appearance prior to the mechanization of design thinking. It is, for example, deeply rooted in ancient Greek thinking. "The shape of heaven must be spherical," argues Aristotle, because outside of this form, "no other part of itself can be found," no "addition" can be made to its surface.[11] In a similar vein, in Hellenistic times the sphere is defined in terms of optimality, as the body that combines the maximum volume requirement with the minimum surface. The formulation of isoperimetral problems by Zenodorus at this time also deals with problems of optimality.[12] In Ptolemy's writings, the circular stellar orbits are seen as optimal, and the same opinion holds in the Middle Ages: "nature and art always strive to arrive at their ends by the shortest means," in the words of Peter of Padua's *Proemium* (1310).

Koyré spoke of such phenomena as part of the general "hantise de la circularite"—the spell of circularity haunting these early scientists in their study of nature. But other features besides the circle also captivated these minds, and again because of their optimality: the straight line, the right angle, the mean, and the center of symmetry.

The obsessive search for patterns of optimization in nature runs parallel to attempts to impose optimal shapes on the manmade world. Writings of the Renaissance, for example, usually prescribe square or circular forms for all types of architectural products, from building elements to whole facades and town plans (Figs. 8–14).

Although these "perfect" shapes are often seen as related to efficiency, as avant-garde occurrences, they are in fact an application of a definite iconographic program: the representation of the perfect, or pure cosmos. This operation of imitation and picturing is subsumed under the archaic form of correspondence of the parts of the universe. In this sense, optimization is a leftover from archaic thinking.

Empirical Verification

The use of accurate standards of measurement and sensitive instruments for recording data is not restricted to modern design practice. Initially, for example, the compass was a geomantic tool for selecting the sites of buildings and graves, and the telescope had a broad astrological application. Similarly, the commitment to empirical verification, to accuracy, and to realism in the representation of the world, although now an indissociable component of the mechanization of design thinking, pre-existed the system proper. Again, we find an archaic predecessor to the modern idea of empirical verification in Aristotle. "Our statements are borne out by the facts," he affirmed, and he criticized those who "in a discussion about phenomena make statements with which the phenomena conflict." [13]

An interesting moment of antagonism between the archaic and the modern use of empirical verification occurred in the early seventeenth century debate between Kepler (1571–1630) and Fludd (1579–1637).[14] Kepler challenged Fludd's contention that the universe was composed of "secret harmonies" that could be made manifest by means of hieroglyphic symbols only. For Kepler, this hermetic model of the world was "vague, ambiguous, obscure, dictated only by the delight in pictures." On the contrary, he held that the order of the world was manifest not in the invisible and the enigmatic, but in what he called the *signatura rerum*, the observable evidence of things. Accordingly, he sought out what could be perceived and accurately mapped in the "geometrical . . . quantitative figures." [15]

Fludd retorted that the harmonies of the world, without which "the whole universe [would] perish and be reduced to nothing," were in the "internal impulses that issued from nature herself" and were therefore invisible to the naked eye. Only "vulgar mathematicians," he accused, concerned themselves with "quantitative shadows." The alchemists and hermetic philosophers, however, comprehend the "true core of natural bodies . . . [Kepler] has hold of the tail, I grasp the head: I perceive the first cause, he the effects." In Fludd's opinion, Kepler was blind to the "first causes" because he was "stuck too fast in the filth and clay of the impossibility of his doctrine." The latter replied: "I hold the tail but I hold it in my hand, you may grasp the head mentally, though only I fear in your dreams."[16]

For all these major differences, the conflict between Fludd and Kepler was one of means and not ends. Both wanted to reveal the harmony of the universe, but one proceeded through divination while the other chose empirical observation and mathematical demonstration as his means.[17]

The possibility of reconciliation was not totally discounted by Kepler, as is evident in his readiness to accept the value of Fludd's work, should it prove fruitful. Indeed, he conceded that if Fludd "found in the very causes harmonies as limpid as mine, then it will be proper for me to congratulate [him] for the gift of invention . . . ."[18]

In a similar manner, the late medieval and Renaissance designers loudly praised the gift of observation and the advances made in the realistic representation of buildings, but they remained devoted to the archaic norm of harmony, and in the same spirit as Kepler's they employed the various modern techniques of accurate measurement and exact representation of line, volume, light, texture, and color. They hoped that the drawing would reveal a secret harmony, or a hidden resemblance with an antique temple or with the Temple of Solomon, for example.

The Analytical Approach

None of the optimally efficient products could have been designed without analytical intellectual tools, without the rigor of the *modus geometricus*. But as much as these modes of analytical thinking were essential for the development of modern efficient products, they existed independently of the norm of efficiency integrated into the conceptual system of archaic design and played a dominant role in the practice of the so-called primitive architecture of tribal societies or sacred architecture of theocratic empires. This is certainly contrary to the nineteenth century romantic interpretation of vernacular building or of pre-Renaissance design as being carried with a freedom of feeling, intuition, and inspiration, an opinion which, although never proven true, has not completely fallen out of favor in our time. Current research reveals increasingly the presence of analytical thinking in North African, Middle Eastern, Indian, and Chinese geomancy and its links with divinatory building and planning (Fig. 15).[19]

Similar historical studies of the lineages of the Pythagorean, platonist, neoplatonist, and hermetic doctrines in medieval and Renaissance architectural theories and practices demonstrate the dominance of quantification in the thinking of the designer. The medieval architect justified design decisions by stating: "What has been done has been done according to geometry:" (*Quod quae sunt, sunt facta per geometriam*).[20] But the aim of geometry was still meant to set up a congruence between manmade product and divine biblical prototype or esoteric cosmologies of antiquity; in other words, geometry was a means of satisfying the norm of purity or harmony.

The idea that all phenomena are measurable and reducible to simple figures never emerged with more conviction than in the debates of fourteenth-century Oxford and Paris on problems such as the measurement of charity in man by theologians like Oresme (Fig. 16)[21] and the Calculator (so named in fact for his influential work, the *Liber Calculationum*, for his obsession with calculations, and for his conviction that everything thinkable was computable).[22] Many of the logico-analytical methods of the Renaissance were developed in answer to questions such as Mersenne's: How was the Ark of Noah able to carry 130 species, how was it possible for the angels to be simultaneously in the sky and on the earth, and how would the valley of Josaphat be able to hold all the dead on final judgement day.[23] In the same vein, but not necessarily with the same intentions, the young Galileo devoted a work to the quantification of Dante's Purgatorio.[24] John Dee, writing as late as 1570 in the introduction to the first English edition to Euclid's *Elements*, saw the ancient philosopher in the light of divination. Dee considered geometrical reasoning to be above all a key for unlocking the mysteries of the divine cosmology.[25] When Jacques Besson in his *Theatre des Instruments Mathématiques et Mécaniques* in 1579 called modern design "fruit of geometry," he expressed archaic preoccupations with the search for secret harmonies as well as modern aspirations for control, measurement, and estimating.

Research in the history of science has shown that if Aristotelian rationalism had governed in the Renaissance along with the idea that it is "on all counts irrational to assign geometrical figures to simple bodies" (which is one of the aspects of the analytical approach), as Aristotle mentions in *The Physics*, neither mechanics nor modern architecture could have developed. Analytical methods and quantification, although essential for the development of modern architectural thinking and mechanization of design, are initially integral parts of the archaic conceptual system of architecture.

Mirabilia and Machines

Like the components of design thinking we have just observed, machines themselves existed during archaic times, prior to the establishment of mechanization. Although Vitruvius, for example, described machines in his *De Architectura*, he did not include efficiency among their primary purposes. Moreover, his use of geometry was limited to analyzing the machine in terms of two elements only, the straight line and the circle, which are employed in a rudimentary way only

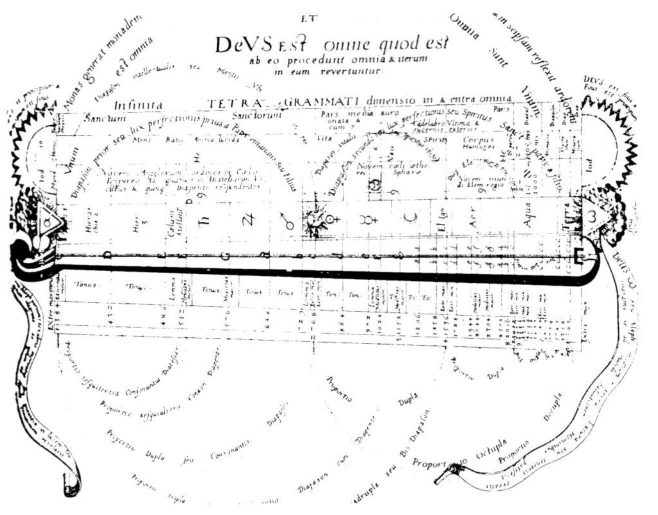
DeVS EST omne quod est
ab eo procedunt omnia & iterum
in eum revertuntur

8

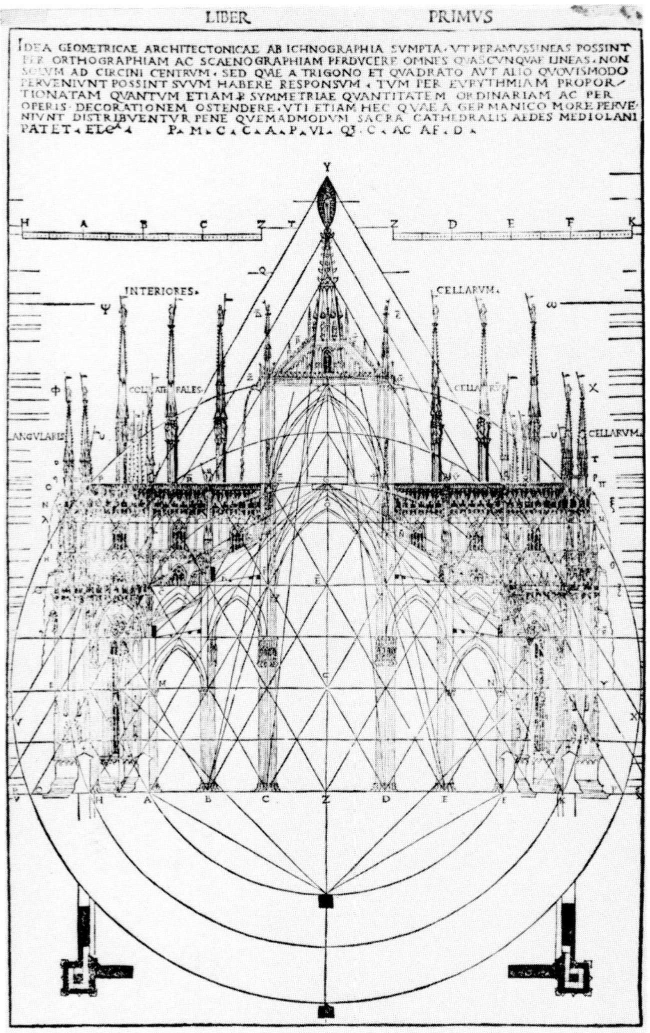
LIBER
PRIMVS
INTERIORES.
CELLARVM.

10

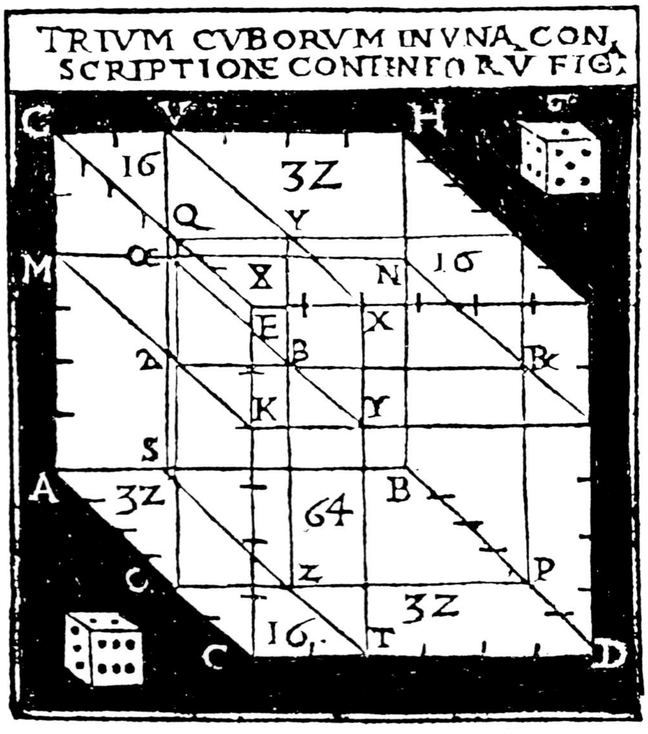
TRIVM CVBORVM IN VNA CON
SCRIPTIONE

9

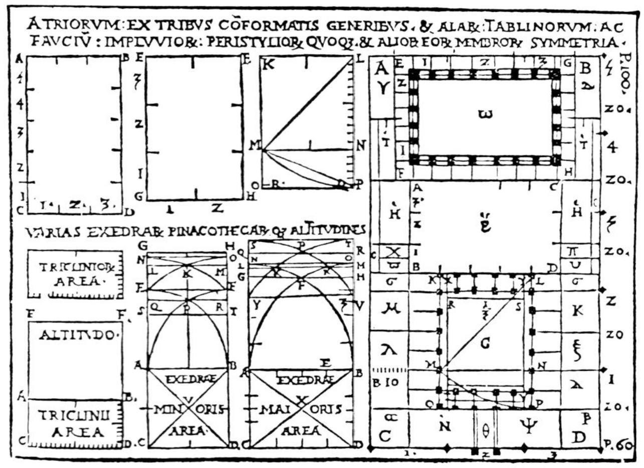
ATRIORVM: EX TRIBVS CŌFORMATIS GENERIBVS. & ALAR: TABLINORVM: AC
VARIAS EXEDRAR: PINACOTHECAR: Q: ALTITVDINES
ALTITVDO

11

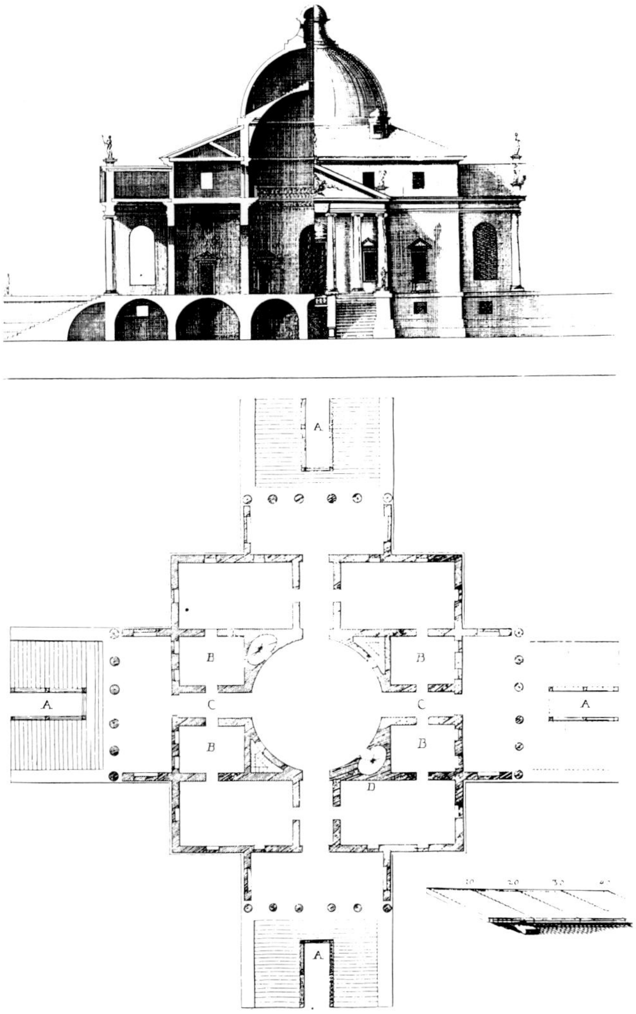

12

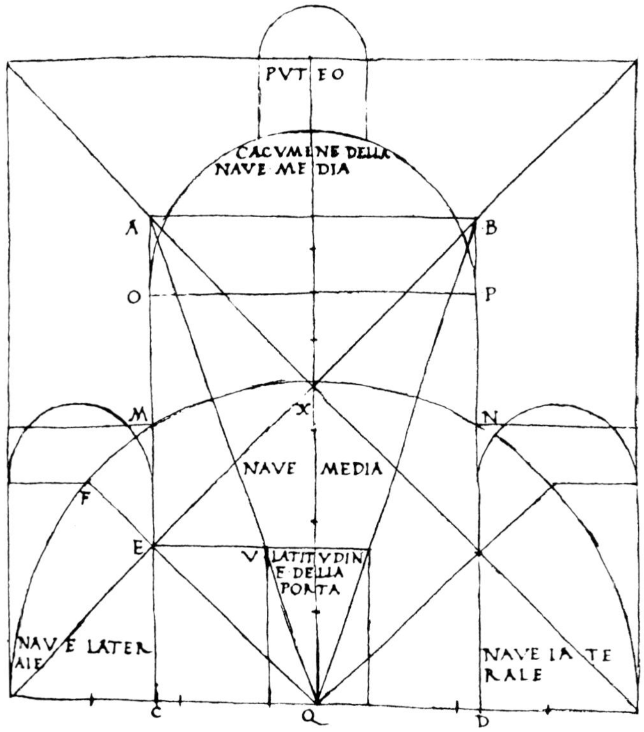

14

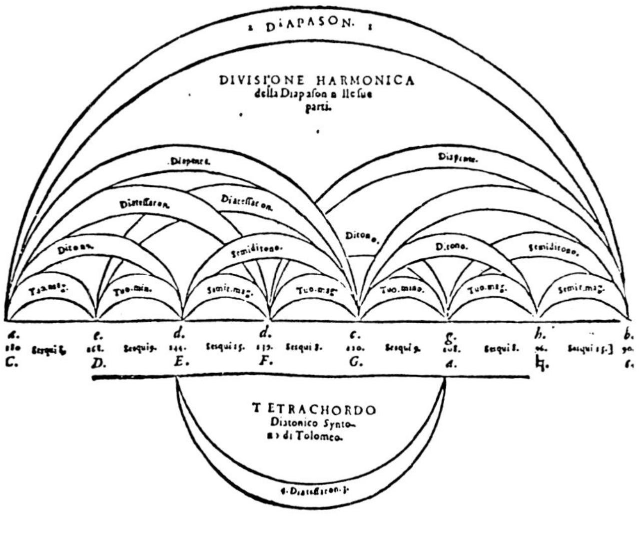

13

8–14. For the architecture of the Renaissance, the norm of purity is obeyed when buildings conform to geometrical relations and proportions believed to be dictated by universal "exemplars" that, in turn, correspond to each other. Fig. 8 shows the monochordum mundi, an analogy between the human body, the divisions of Plato's heptachord, and the tabernacle of Moses. Fig. 9 shows a system of harmonic divisions of the cube from Cesariano's commentaries on Vitruvius, De architectura *(Como: 1521). Fig. 10 gives the same analysis applied to the Cathedral of Milan, and Fig. 11, the same applied to an atrium house. Fig. 12 shows the Temple of Solomon as drawn by the Jesuit Villalpando in his* In Ezechielem Explicationes *(Rome: 1596–1605). Fig. 13 is a diagram of a set of proportion rules abstracted from any concrete object, found in C. Zarlino's* Institutioni Harmonische *(1573), a book widely read by architects during this period. Fig. 14 shows the section of a church from Philibert de l'Ormo's* Nouvelles Inventions *(Lyon: 1561); such a section was still determined by the use of archaic rules of geometric relations and proportions still obeying the norm of purity.*

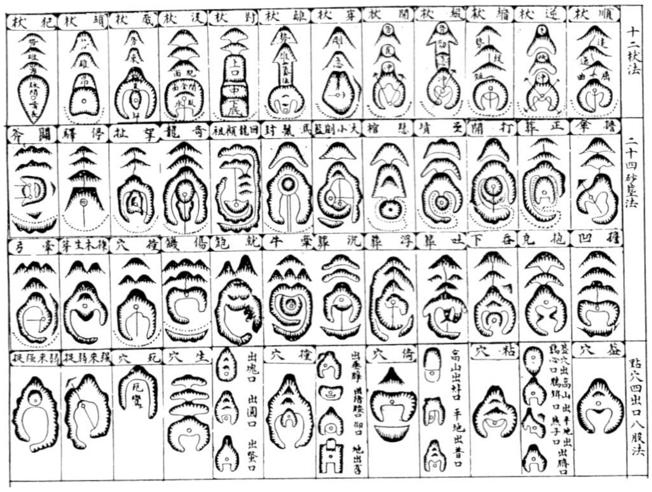

15. A page from a Feng Shui manual that describes an archaic analytical method used in China to identify the location of a building. The illustration shows a classification system (according to Yang Yun Sung) of twelve lineages, twenty-four sand formations, pits, and openings with auspicious and ominous locations.

16. From Nicolas Oresme, The First Graphic Representation of Change, c. 1350, in his Tractus de Latitudine Formarum, *Second edition (Padua: 1486).*

to describe the machine, not to explain how it is made. Empirical observation was hardly used with regard to the organization of the machine. What we do find however, is that Vitruvius seems to tie the machine to the archaic doctrine of correspondence between artificial objects and the astral macrocosm (X,2), judging from his statement that "mechanics is founded . . . on the study of what causes the circular movements of the cosmos."

A closer look at Hero of Alexandria's machines reveals that they too were unrelated to concerns of efficiency. These machines belong in fact to what we may call the realm of "mathematical magic"; they were, like their sixteenth-century counterparts, "thaumaturgic," in other words, machines whose aim was to overwhelm the spectator with awe in the contemplation of the hidden powers at work in the world. Or again, these machines fell under the category of "mirabilia" or playful objects, and it is from these that a seventeenth-century designer, Salomon de Caux, drew his inspiration for his famous garden machines, with which he intended to surprise, delight, tease, and amuse, and which we find illustrated in his *Les Raisons des Forces Mouvantes* (1615).

The norm of efficiency appears to be linked to machines for the first time in the restricted field of the military. Why the idea of economic gain through machines was not more widespread in antiquity has been the subject of great dispute. The socioeconomic context, the constraints of the slave economy were probably the major cause.[26] It has been argued that within this system only the increase in the number of slaves could improve productivity. Since the foremost means of obtaining slaves was warfare, and since the means for carrying out warfare were limited, all efforts were centered on making it efficient. If there was an economic interest attached to military machines it was only accomplished indirectly, by insuring victory in battles and consequently bringing in slaves. This explains why the catalytic role of the machine in the development of mechanization should have occurred in connection with military engines, military products, military sites, and military engineers (Fig. 17–18).

With the decline of the slave-based economy and the inward-looking early feudal economies and with the rise of the market, another field of economic gain was revealed, outside of military operations: the control of the processes of production aimed at efficiency.

Machines and Mechanization: The Contribution of Galileo

In pre-Renaissance writings machines were not conceived differently from other objects. They fit with all other artificial objects into the same framework of archaic thinking. By the end of the sixteenth century the design of military machines was viewed as a violation of archaic design norms.[27] This occurred because the design of the military machine finally had become associated with a new type of thinking whose elements had been dislodged from their original mold and combined in a novel way.

In the beginning of the seventeenth century we see the concepts and methods we referred to before as separate entities assembled in the design of the military machine. The association between them became so strong that the machine seems for some time to be the only object capable of embodying the new thinking. Eventually the thinking was abstracted from its catalyst; once this abstraction occurred, the new type could be generalized and projected into the design of any product, taking the place of the archaic doctrines of architecture.

Machines, like the components of mechanized

design thinking, existed prior to the rise of the market, the logic of a monetary economy, the mercantilist mentality, and the interests and aspirations of the new commercial class that began to dominate society and control production in the Renaissance; they existed prior to these new realities but they also rose to meet them when they began to appear and mesh with them. Machines and mechanized thinking, although not inherently modern, were absorbed into what Freud called the bourgeois "reality principle," which drew on it whatever it could find. Sometimes, this absorption was not without resistance. Pico della Mirandola, for one, argued that "the science of numbering, chief and most divine . . . should not be confused . . . with the mercantile arithmetic."[28] So, as much as the practical and intellectual tools of mechanization belonged to the *ordo mercantorum* of the late Middle Ages and Renaissance, to its love of action and business, of money and bookkeeping, bills of exhange, and pricing strategies, we also find these tools at home with the archaic ideals of contemplation and purity. One may, however, suggest the reverse, that although these tools of design developed initially within what we have called the archaic conceptual system of architecture, they were a necessary precondition for the emergence of modern design thinking.

Idea versus Matter

Archaic design thinking depends on facts that are supported by backings that were provided by sacred cosmologies, divination manuals, and theological writings containing taxonomies of pure and impure objects. These taxonomies were validated through references internal to the accepted beliefs that made up the tribal myth, which acted as the base in the archaic conceptual system. Reference to the base is a form of validation, mainly from the view of logical consistency, of whether or not a given statement confirms or does not confirm a pattern of universal correspondence or harmony.

During the period of transition from the archaic world order to the modern, that is, during the late Middle Ages and the Renaissance, backings and bases in the conceptual system of design were provided by a derivative of archaic thinking: the neoplatonic doctrine of Idea.[29] Drawing from the writings of Plato and Plotinus, but also from beliefs widespread throughout the archaic world of the Mediterranean, the doctrine asserted that the design product must embody Idea that exists in a "pre-figured" way prior to the product and independently of it; as Ficino argued, "the architect's first notion of the building, *ratio*, dwells in his psyche in the form of Idea." Architecture was believed to capture this perfect essence, this *forma exemplar*, as a mirror captures an image, to use Plotinus's analogy. Thus, as Pico della Mirandola claims, for every building there are two images, one "ideal" in the spirit, the other, "of the senses in the marble or

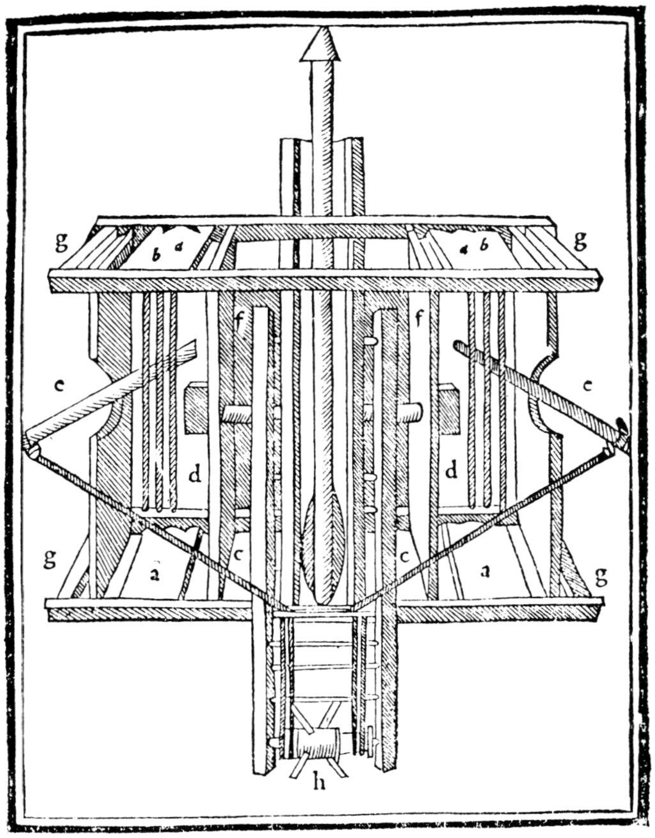

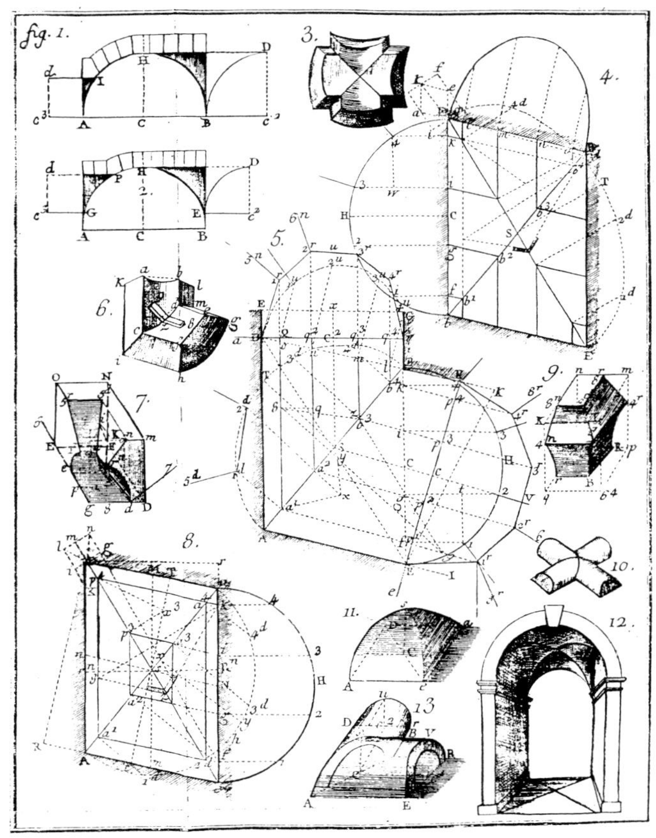

17–18. Roman military engines are shown in Fig. 17, an illustration from Jean Martin's pre-Galilean edition of Vitruvius (1547). Fig. 18 shows diagrams for vault construction that contrast the two modes of graphic representation. From one of the first books on engineering, A. F. Frezier's La theorie et al practique de la coupe des pierres *(Paris: 1745–59).*

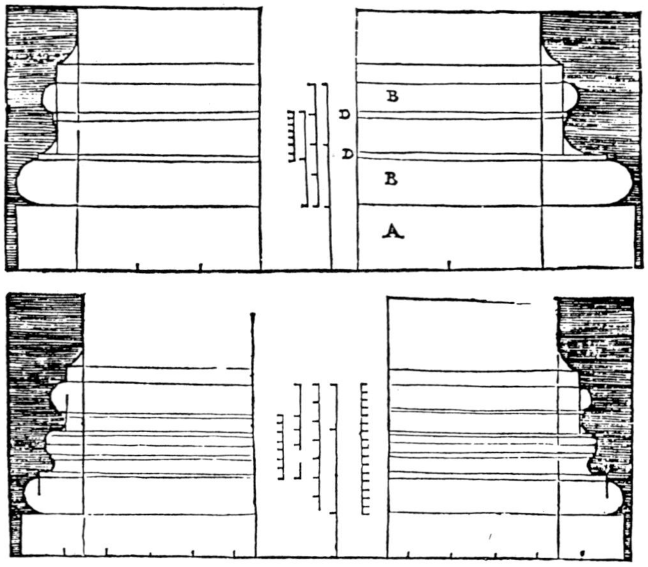

19–20. Two illustrations from Alberti. Fig. 19 showing the Doric base and Fig. 20 the Ionic base. In both cases, the figuration stresses the geometry of the profile and the proportions, disregarding size and material.

stone or other material."

We can see how the doctrine fit into the framework of archaic design: matter shaped into a design product followed from Idea, which was nothing but the set of rules for the mixing of pure or harmonious objects. The more the product resembled the Idea, the more pollution was avoided, the more harmony is established and the more the building became an *opus perfectum*, a perfect artificial product.

With the mechanization of architecture the position was overturned. The form of the product was determined by its material essence, the stuff of which it was made and its size. The idea of a design was nothing but the pattern of the physical object constrained by its natural attributes.

The break between archaic and modern conceptual systems of architecture at the level of fact, backing, and base is best illustrated in Galileo's views on the relation between the form of a product and its matter around the end of the sixteenth century.

In his *Discorsi e dimonstratione matematiche intorno a due nuove scienze*, written in spite of a prohibition by the Church and subsequently published by Elzevier at Leiden in 1638, Galileo took the decisive step in the conceptual revolution that eventually shattered the archaic framework of design. This change occurs at the level of fact, backing, and base.

Galileo acknowledged a debt to the Venetian technicians for the "constant activity which they display in their famous arsenal." He singles out those who "partly by inherited experience and partly by their own observations, have become highly expert and clever in explanation."[30]

The activities and especially the theories of these "first rank men" formed the subject matter of the opening pages of the *Discorsi*. Here Galileo expressed special admiration when one technician explained that the reason why he and his co-workers used "stocks, scaffolding, and bracings of larger dimension for launching a big vessel than they [did] for a small one . . . in order to avoid the danger of the ship parting under its own weight" was that "many devices which succeed on a small scale do not work on a large scale."

This statement about scale is similar to an observation that had been made earlier by Leonardo da Vinci (Manuscript A, fol. 3V and fol. 46V) on the equivalence of pillars on struts and on beams. Leonardo remarked that "a support with twice the diameter will carry eight times as much weight as another, both having the same height" and that "many little supports held together are capable of bearing a greater load than if they were separated from each other." But he neither characterized this as exceptional in relation to contemporary doctrines of design nor attempted to generalize such observations.

By contrast, Galileo focused on the event as an anomaly with broad implications. He saw it as putting into question the archaic doctrine of the primacy of Idea over Matter, which held that if a large object is "constructed in such a way that its parts bear to one another the same ratio as in a smaller one, and if the similar is sufficiently strong for the purpose for which it was designed," then "the larger also should be able to withstand any severe and destructive tests to which it may be subjected."

He saw an even more serious challenge to the archaic doctrine according to which "mere size cuts no figure . . . the properties of circles, triangles, cylinders, cones, and other solid figures [do not] change with size."

Indeed, with the advent of Galilean mechanics, proportion was no longer governed by the perfect "Forms" of the world, by its divine "Idea." It was seen rather as something bound to matter that changes according to the size and material of an object.

Contrast these views of Galileo's with Alberti's, who two hundred years earlier had written

> We can in our Thought and Imagination contrive perfect Forms of Buildings entirely separate from Matter by settling and regulating in a certain Order the Disposition and Conjunction of the Lines and Angles. Which being granted, we shall call the Design a firm and graceful pre-ordering of the Lines and Angles, conceived in the Mind, and contrived by an ingenious Artist.[31]

A similar contrast emerged in the justification of the use of building models by Alberti. He suggested the use of "real Models of Wood or other Substance." These models were not for "charming the Eye," but to examine, in addition to aspects of "ornament," "how the Walls are made, and how strong and firm the Covering." The model therefore was explicitly applied to test "Strength," without any awareness of problems that would emerge out of the reduction of size in the simulation (Fig. 19–20).

Furthermore, by referring to the "deviations observed between the machines in the concrete and in the abstract," Alberti upset the belief that the former had to obey the latter. On the contrary, for Galileo the "abstract," the Idea, were reflections of the "concrete," and the Matter and became subservient to the size and material of products.

With Galileo the ancient formulae were broken, the system of design overturned. "The mere fact that it is matter" frees the machine from the hold of invariant, archaic "correspondences." At the same time, however, the machine was subjugated to a new rule, tied to the changing material conditions, conditions which could be known only through "observation" and "demonstrated by geometrical methods." If a large machine is built "of the same material and in the same proportion" as a smaller one, then the larger "will not be so strong or so resistant against violent treatment." Proportions and form alone could not keep it from breaking.

The initial observations made by Galileo on machines are then transferred to construction elements such as pillars, cantilevers, and architraves. Thus, gradually the complete building fabric was seen in analogy with the machine. Its form did not follow the Idea; instead it was determined by the behavior of matter. The shape of its components depended on their material, their composition, and their size.

What held for machines also held for buildings. Similarly, what held for buildings held for any natural body as long as it was seen as a construction. The form of a skeleton of a horse, a dog, a cat, a grasshopper, as well as an ant, were examples invoked by Galileo to point to the dependence of form on size and material generalizing mechanics as a theory of construction applicable to any fabric. The justification of the form of the machines becomes the explanation of the form of nature (Fig. 21):

> To illustrate briefly, I have sketched a bone whose natural length has been increased three times and whose thickness has been multiplied until, for a correspondingly large animal, it would perform the same function which the small bone performs for its small animal. From the figures here shown you can see how out of proportion the enlarged bone appears. Clearly then, if one wished to maintain in a great giant, the same proportion of limb as found in an ordinary man, he must either find a harder and stronger material for making the bones, or he must admit a diminution of strength in comparison with men of medium stature; for if his weight be increased inordinately, he will fall and be crushed under his own weight. Whereas if the size of the body is diminished, the strength of that body is not diminished in the same proportion; indeed the smaller the body the greater its relative strength. Thus a small dog could probably carry on his back two or three dogs of his own size; but I believe that a horse could not carry even one of his own size.

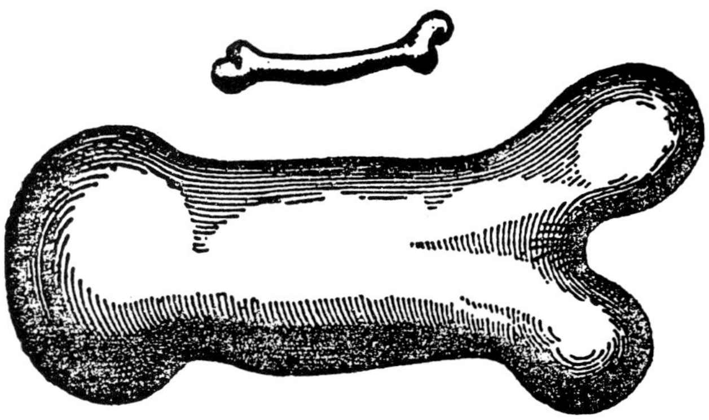

21. Illustration from Galileo's Due nuove scienze *demonstrating the doctrine of the invariance of proportions to be false, as invalid in the natural world as in the world of design. This is one of the first examples of the mechanization of the conception of natural objects and is also an early instance of biometry.*

The aim of Galileo's innovation was not only to upset the belief that placed the "Idea" on a higher plane than matter, and myth above observation and empirical data. His main objective was efficiency in building. The commitment to efficiency was presupposed within the notion of the "limit," absolute resistance in the propositions of the *Two New Sciences*: that, "for every machine and structure, whether artificial or natural, there is set a necessary limit beyond which neither art nor nature can pass," that there is "one and one only" structural element which, when loaded, performs to the "limit between breaking, and not breaking" while any other would either break or be "able to withstand some additional force tending to break it," being redundantly strong.

The notion of the limit brings us to a new definition for monstrosity and ugliness. In the typical archaic manner, Francesco Giorgi, in his *Memorandum for S. Francesco della Vigna* of April 1535, stated that one would create a "monster" by breaking and violating the "natural laws." He allowed for increases in measures and numbers, as long as these retained the original ratios. Galileo, on the contrary, held that monsters in nature result from ratios being kept constant while measures and numbers increased.

> You can plainly see the impossibility of increasing the size of structures to vast dimensions either in art or in nature; likewise the impossibility of building ships, palaces or

22. *Illustration from a Dutch edition of the works of Vauban,* De l'attaque et de la défense des places *(Hague: 1737). It represents "all the pieces which make up military architecture or the art of fortification." The rational combination of these simple elements could solve any fortification problem, it was believed, since Vauban held that the first principles and definitions of Euclid could yield the whole universe of geometrical theorems. The polygonal bastions on the right represent one of the first attempts to create an optimal (mini-max) plan. In the words of Vauban, "Les fortifications d'une place peuvent estre appelées parfaites, quand elles fournissent a un petit nombre de soldats toutes les commodités imaginables, pour resister avant ageusement aux efforts d'un puissant ennemi qui vent les forcer."*

> temples of enormous size in such a way that their oars, yards, beams, iron-bolts, and in short, all their other parts will hold together. Nor can nature produce trees of extraordinary size because the branches would break down under their own weight; so also it would be impossible to build up the bony structure of men, horses or other animals so as to hold together and perform their normal functions if these animals were to be increased enormously in height: for this increase in height can be accomplished only by employing a material which is harder and stronger than usual or by enlarging the size of the bones, thus changing their shape until the form and appearance of the animals suggests a monstrosity . . . ."

This was a new definition of monstrosity, indicating a departure from archaic doctrines that had seen in deformation an evidence of pollution. For Galileo the explanation entails no sacrilege.[32]

Mechanization of the Building Plan and the Paradigmatic Use of the Machine

The machine eventually came to be identified as an anomaly in relation to archaic thinking and was subsequently used to undermine the dominance of archaic thinking over architecture. The machine also played the role of a conceptual catalyst to bring together isolated, preexisting elements into a new conceptual system. To these two methodoligical or heuristic uses of the machine in development of modern architecture was added a third, what we might call the paradigmatic use of the machine.

The machine as a paradigm became a shorthand description of the new conceptual system of design. When a designer would say that a building was a "machine," this meant that the concepts and methods to be followed in the design of a building were contained in mechanics. In addition, behind this brief description of the new conceptual system as a machine was a rhetorical device which made the generalization of the new philosophy of construction legitimate. It made the application of mechanics appear through an argument by analogy as desirable and permissible to other objects besides machines.[33] It was an invitation to make buildings appear removed from the normative network that linked them in the archaic world with the divine bodies of the astral orbits, the musical instruments, the human body; it attached buildings to the new world order of the market and of effective production.

Initially, the emergence of a new conceptual system impressed very few. Indicative are Galileo's politico-ideological troubles; as Giorgio de Santillana remarked, "Roman society hardly took notice of the whole affair . . . There were more absorbing subjects to talk about, in that summer, 1633, such as Bernini's enormous and magnificent baldachin with the twisted columns which had just been inaugurated in St. Peters." About a century later the situation changed. When cracks appeared on St. Peter's dome it was three "mathematicians," Le Sueur, Jacquier, and Boscowich, men who were attached to the framework of thinking, who were invited to report on the causes of the damage and recommend solutions, not architects.

The spread of the mechanization of architecture was slow but irrevocable. We have the theoretical efforts of Lodoli around the end of the seventeenth century to create a new system of architectural thinking based on the achievements of mechanics. As Andrea Memmo's *Architettura Lodoliana* (1786) reveals in the published outline of Lodoli's book, now lost, the scheme was extremely ambitious and perfectly clear in its intentions. The ground to be covered was enormous, however, and the endeavor remained at the level of generalities. We have also the most stimulating writings of Bartolomeo Vanni, more critical than constructive in relation to the state of architecture. Thus, it is in concrete problems of design that we should turn to find the first instances of the spread of mechanization.

By the end of the eighteenth century mechanization of design spread beyond the problems of construction and stability into the plan of the building and its operation. The new thinking was transferred to deal with the efficient performance of the building as an envelope of activities and processes.

Here again one must turn to military architecture for first developments. "The force of the fortress depends more on the quality of the plan than on the thickness of the walls" noted Francesco di Giorgio, underlining the superior performance in defense of the triangular bastion over the circular tower (Fig. 22).[34] A clear case of mini-max formulation emerged in the sixteenth-century fortifications, where a compromise was sought between the bluntness of the bastion with the length of the curtain.[35] During the seventeenth and eighteenth centuries the projection of the machine as a paradigm into the plan of fortifications became more evident as the designer tried to find solutions that achieved maximum security for the minimum number of cannons and operators, maximum defense for the minimum expense in construction.

By the end of the eighteenth century the numerous projects by the French Academy of Science for the Hôtel Dieu in Paris[36] and the elaborate calculations of Bentham for the multipurpose controversial Panopticon buildings[37] revealed that mechanics had extended beyond the material fabric into problems of space allocation and arrangement of activities (Fig. 23–25). As in mechanics the search was for a structure which, without collapsing and for less material, creates more usable space; in the case of these welfare architecture plans, the designer tried to find a plan by which the maximum service was achieved through

23–25. Two early examples of buildings conceived as machines "to cure patients" are described in proposals for a hospital in Paris to replace the ancient Hôtel Dieu. Both proposals are based on the idea of the building as an air pump. Contemporary ideas about medicine are combined with the state of development of air dynamics. The idea of the plan is also mechanistic, placing services in the center of gravity of a building, a central point in Fig. 23, A. Petit's Memoire . . . de constuire un Hôpital . . . . *(Paris: 1774) and a line in Fig. 24, J. B. LeRoy's* Précis d'un Ouvrage sur les Hopitaux *(Paris: 1787). Patients are assigned a minimum space defined again through mechanistic studies which deserve to be seen as one of the earliest examples of* existenzminimum. *Fig. 25 is a similar case of a building conceived as a machine from Jeremy Bentham's* Panopticon *(1787). In all of these cases, not only is the nonessential ornament eliminated, but the plan and facade of the building no longer derive from the traditional rules of composition and ordering.*

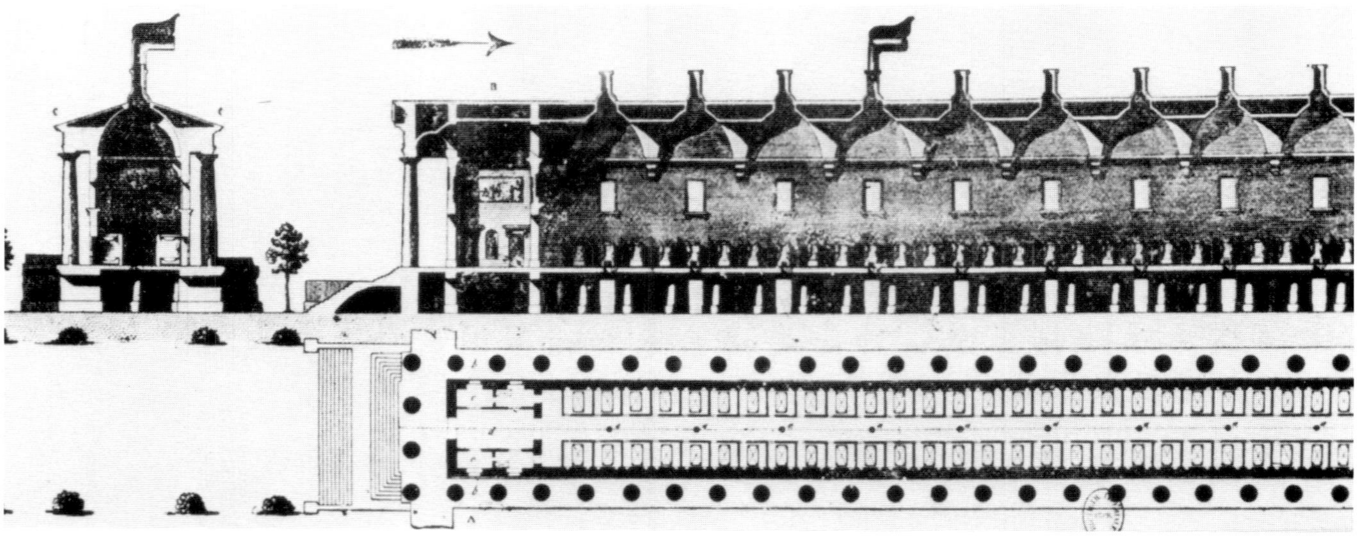

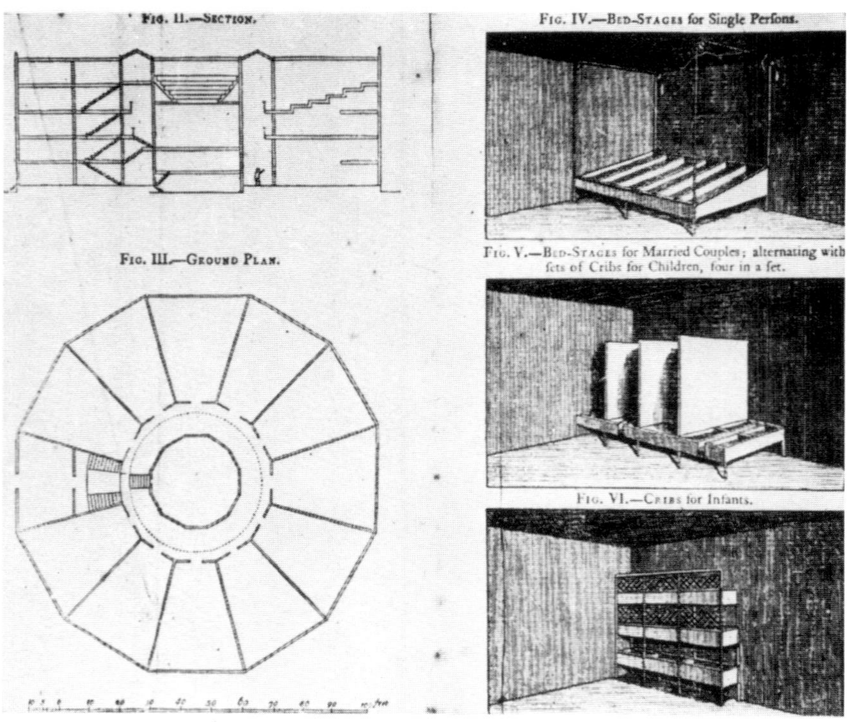

the smallest number of people. Problems of location of such service buildings were discussed for the first time also across the same lines of optimization.

The spreading of mechanization of architecture was the result of a combination of conceptual developments and of the emergence of new interests. It was also the outcome of the establishment of new institutions. Intellectual innovations could be completely ineffective if not accompanied by institutional changes. Rarely in history did arguments themselves provide sufficient grounds for the adoption of the new thinking framework possible. This was the case with mechanization of architecture. The establishment of a number of institutions in Europe indeed helped diffuse the mechanization of design and in turn accelerated the creation of new knowledge within the conceptual framework. Among these institutions were the Académie Royale d'Architecture, the Académie des Sciences (Fig. 26), the Jardin du Roi, the army, and especially, the Ecole des Ponts et Chaussées and the Ecole Militaire de Mezieres.

Under instructions from Colbert, the Académie Royale d'Architecture undertook in 1678 the extensive study of building materials. The academicians observed in a systemic and detailed way the quality of stones used in construction around and within Paris and studied the role of the sun, wind, the rain, and even the moon on the aging and corruption.[38] Other studies followed: strength tests of wood, metal, and glass rods by Mariotte (1670–80), bending tests on oak wood beams by Parent (1707–08), the measurement of the sagging prior to rupture by Buffon. These were some of the first steps, inconceivable without institutional guidance and support, that later culminated in such projects as the first bridge over the Seine at Neuilly (1768–1772), whose pier size was determined through strict calculations stemming from mechanics, and the Church of Sainte-Genevieve in Paris, today's Pantheon, the first building whose column dimensions were dictated by rigorous application of mechanics.[39]

Mechanization and the Aesthetic Approach

Mechanization was not the only impact of the market, however. Another approach to architecture emerged simultaneously. Its concerns were questions of beauty. This approach was opposed not only to archaic thinking but also to the mechanization of architecture. The norm that governed it was neither sacred harmony nor economic efficiency, but aesthetic gratification. Aesthetic gratification meant the search for pleasure in the visual organization of design products and was characterized by fantasy, a distaste for analytical methods, a contempt for empirically observed data, and an impatience with logical procedure. The aesthetic approach seemed to be if not hostile, at least indifferent to mechanization. On a deeper level, it was complementary to it.

26. Louis XIV visiting the Academy of Science. Note the framed plan of the bastion capturing the attention of Colbert and Louis, and in the background, the observatory of Claude Perreault under construction. The study of these buildings was paralleled by the design of instruments and the investigation of the form of natural objects, both of which abound in the illustration from Mémoires pour servir à l'Histoire Naturelle des Animaux *by Perreault (Paris: 1671).*

To a great extent the growth of the mechanization of design, as we have already mentioned, followed from a deeper interest in economic gain, after the market emerged as the prime mover and organizer of human relations. The control of production costs became a major preoccupation. It eliminated the archaic fusion of wish and accomplishment by pressing the designer to pin down his aims through logic, observation, measurement, and testing, by prohibiting him from dreaming away his plans.

Aesthetization of design, on the other hand, encouraged fantasy; it played with make-believe and placed the user of the environment in "as if" situations, although in a different way from archaic design. Its opposition to mechanization was not regressive. It reacted to an interest equally new, equally linked with the emergence of the market, but its direction was different: the maximization of consumption. Its aim was to make the design product an object of desire, to turn buildings, cities, and furniture into an intricate matrix that titillated the senses.

It also helped, through the dressing up of buildings and cities in the manner of antique times perceived as heroic, to provide an image of continuity,

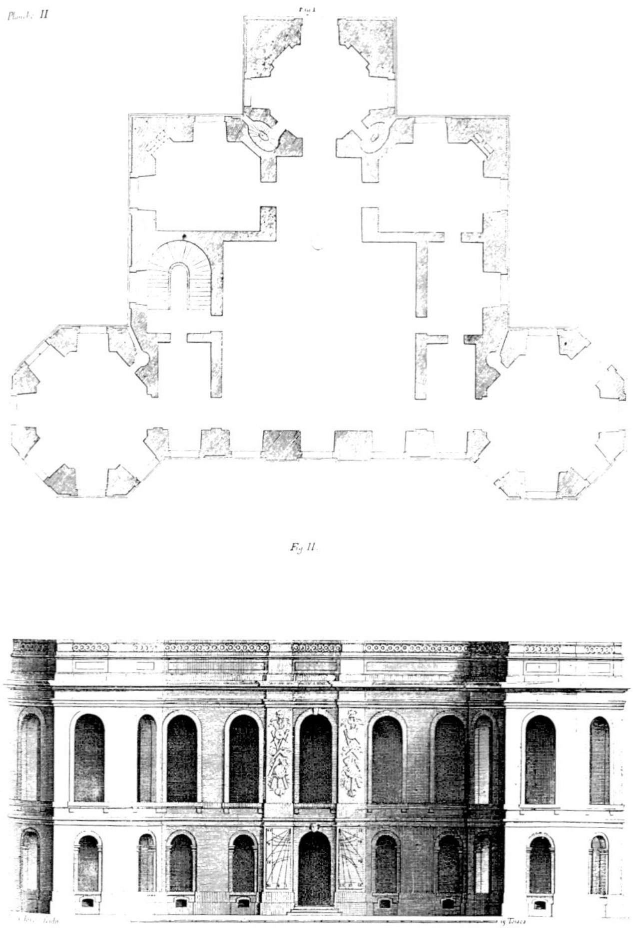

27. *Claude Perrault, Paris Observatory from* Les dix livres d'architecture *(Paris: 1673).*

stability, respectability, and legitimacy to the absolutist state, a new political order protective of the still fragile network of the growing commercial transactions.

Both the aesthetization and the mechanization of design followed from the same new social order and the same new culture. Like pistons in a working engine, their action was complementary. They pumped life into the same social body and they were interdependent. Yet, by the same token, they were antagonistic. This opposition reflected a conflict that was being carried out, again, in the social order of the market.

The Building as an Instrument and as an Arbitrary Beauty

The attack on Claude Perrault's design for the Paris Observatory (Fig. 27)[40] by Cassini, an Italian astronomer and its first director, illustrates how opposed the two tendencies of aesthetization and mechanization were. Cassini complained that Perrault had been too preoccupied with the orders and with their correct, beautiful, and majestic arrangement, concerns typical of the architects. For Cassini, a scientist, these concerns were unacceptable because, in his words, "even the building is an instrument,"[41] meaning that it had to be designed with efficiency above all in mind.

Now Perrault was anything but a stranger to the type of thinking that held efficiency as its highest value, to mechanics. He was a member of the Académie des Sciences, a friend and collaborator of Huygens, and a specialist in the construction of machines and he even applied the model of the machine to the human body in his physiological studies and treatises. What made such an expert momentarily discard his knowledge?

The answer lies in his famous essay *L'Ordonnance des Cinq Espèces de Colonnes* (1682). There he explained that the beauty of architecture was "arbitrary," that its laws, in other words, were attached to the shifting values of society and were of a different order from the "positive" or absolute laws observed in mechanics.[42]

Perrault's design, then, came as a result of a conscious choice of beauty over efficiency. He was a theoretician and a practitioner equally at ease with the approaches of aesthetics and mechanics. That his design was an exercise in one at the expense of the other reveals how incompatible beauty and efficiency were perceived to be.

As the techniques of mechanics steadily developed, the ideas of aesthetics advanced too. It became harder and harder for the mechanical side of design to share notions, expressions, idioms, heroes, prototypes, and even institutional settings with the aesthetic tendency. Belidor, a prominent engineer and builder of a large number of bridges and other large-scale projects at the beginning of the eighteenth century, found it inconceivable how architects operated without any knowledge of what the engineer considered fundamental construction tools;[43] meanwhile Boullée, a prominent designer of projects, mostly unbuilt, stressed the point that architecture had nothing in common with construction.[44]

The emergence of a new way of thinking in design is always accompanied by the generation of new institutions to support it, and the appearance of two opposed conceptual systems towards the beginning of the eighteenth century was paralleled by institutional antagonism.

We have already referred to the role of the Royal Academy of Architecture in the mechanization of design during the last part of the seventeenth century. But equal in importance were its discussions on aesthetics. The dual role which the Academy was initially meant to fill no longer existed by the mid-eighteenth century. It was then that the newly founded Ecole des Ponts et Chaussées drained the Academy of the most important part of its technical teaching. By the end of the eighteenth century, two entirely new institutions—the Ecole des Beaux-Arts and the Ecole Polytechnique—took the place of the old Academy, each charting its own direction, each developing its own curriculum, its own style of teaching and its own image of design in the minds of its students. At the Ecole

Polytechnique, mechanics was purified, sharpened, and carried to the extreme heights of excellence and specialization.[45] At the Ecole des Beaux-Arts, aesthetics found equal success and became a powerful design method, firing the design product with imagination, desire, and flamboyance.

The Functionalist Antidote

The conceptual and institutional division between mechanization and aesthetics in design was the source of conflicts, confusion, and redundancy. Architecture was declared to be in a state of crisis, and that statement has been incessantly repeated ever since. The need to arrest the negative aspects of this split triggered attempts at reconciliation.

The lack of a common conceptual framework between engineer and architect was seen as the cause of losses in the construction and planning of buildings and larger design projects. Campaigns originating on the side of the state or other institutions representing mostly collective social interests tried to reconcile the two approaches.

Such efforts were usually carried out by designers moved by a mixture of pragmatism and vague idealism. César Dali and his monumental work, the *Revue General de l'Architecture et des Travaux Publics* offers one of the earliest and most prominent examples of this reforming spirit aimed at overcoming, for the sake of a more socially and economically rational building profession, the growing divergence between "architects and engineers . . . perfectly isolated from each other."[46]

As early as 1699, J. F. Félibien makes a distinction between apparent and actual firmness of a building. In his *Dissertation touchant l'Architecture Antique et l'Architecture Gothique* he remarks that "it is not enough that a building be constructed solidly, it must appear so to the eye" (Fig. 28) and Frémin in his *Mémoires* of 1702 states clearly that buildings have to signify their construction.[47] Gradually, the visual rules, the style, started to appear, fulfilling the functionalist program. The polemical writings of Laugier (Fig. 29–30) and Milizia were intermediary steps to what is to become the functionalist style. This, through the manipulation of a limited number of architectural elements, presented saving as consumption, and consumption as part of the process of production. Ornaments might be stripped from the surfaces of the building, but the building itself became an ornament.

Straight lines, right angles, and the grid patterns cutting the spaces of the building with a jeweler's delight gave the appearance of an analytical rationality in construction and operation. The free curves and regular irregularity seemed to grow organically out of the great machine of nature. Cubes, prisms, and cylinders were used in parceling space to make it look as if it were an assemblage of machine parts. Colored, polished, chopped and tinted plaster and plywood served as camouflage devices that concealed masonry walls and plain carpentry in order to imitate machinemade materials (Fig. 31–32).

28. From Castelli e Ponti (1743) by N. Zabalia, an engineer mainly of scaffoldings and one of the greatest early exponents of applied mechanics. Note in the illustration characteristics of early rigorist functionalism. All ornaments are eliminated. There is an emphasis on the colossal and the transparent as a result of engineering know-how. Contrast these attributes of the new building with the more traditional buildings in the background.

What was needed was a new, broader conceptual framework to make mechanization and aesthetization compatible. One effort went into developing a psychology of sensory gratification. It projected the paradigm of the machine onto the psyche. As a machine reacts to external forces, so the psyche was perceived as responding to external visual stimuli: lines, planes, volumes, and colors. This is when what we today call environmental determinism and scientistic aesthetics emerged.

But the effort was not very successful. First, the evidence was weak, not to say contradictory, from the beginning. As Claude Perrault had already demon-

29–30. Fig. 29 is Laugier's frontispiece from his Essai (1755) in which the primitive hut is proposed as the model for rational architecture made up of only "essential" parts. Although not very original in its ideas, Laugier's text stirred great interest and succeeded in being seen as the most representative of the early manifestoes of functionalist architecture. Fig. 30 is another example of early functionalist organic architecture from a lesser known but in many respects more original work than Laugier's, Ribart de Chamoust's L'Ordre Francais trouve dans la Nature *(Paris: 1783).*

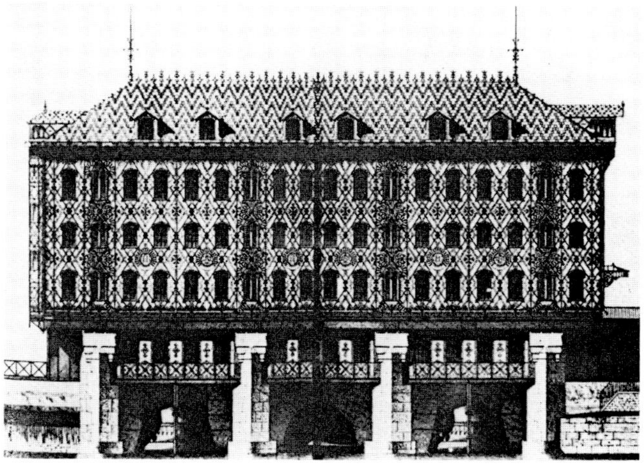

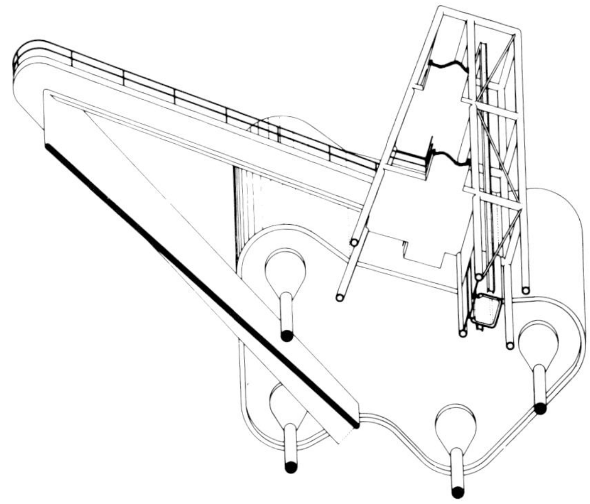

31–32. Two contrasting ways of resolving the crisis that results from the mechanization of architecture. Fig. 31 shows the Mernier Chocolate Factory by J. Saulnier, 1871. The facade reflects the conceptual dichotomy between minimization of production costs and maximization of consumption pleasure. This dichotomy leads to the split of architectural tissue into bearing screen of steel and a decorative "tapestry" of colored glazed tile. Fig. 32 shows that, in James Stirling's project for the Olivetti Headquarters in Milton Keynes, the entire building plan is conceived as a machine-like ornament.

strated in the seventeenth century, historical data did not support the machine-psyche analogy. Neither did the anthropological data that started pouring in during the eighteenth century. The analogy itself did not lead to the desirable conclusion, which was that mechanization and aesthetization were compatible because they could yield the same results, the same design pattern.

Therefore, in the midst of acute credibility troubles for architecture, in the absence of a more theoretical accommodation, a new doctrine was born—Functionalism—asserting the reconciliation of the conceptual frameworks of mechanization and aesthetization. In this it relied more on dogmatic insistence than on rational discourse.

The word functionalism means many things to many people. It has been used often in reference to the bias of some modern designers in favor of utilitarian, economic or scientific considerations, against ornamental or symbolic concerns. It has been linked to the moralistic critique of the fabulatory uses of architecture, calling for an architecture that stands for honesty and projects the aspirations of an ethical world. Functionalism has also been equated with the doctrine that tried to resolve the contradiction between mechanization and aesthetization, and in doing so created a new visual organization of buildings, a new style containing an implicit social and cultural message.

The latter definition concerns us here. The message of the functionalist style was that the norms of maximization of pleasure in consumption and the minimization of cost of production did not need to lead to contradictory behavior, that the act of production did not have to constrain and that consumption did not diminish the potential for production. In a realm of make-believe worked out with the iconography of the building, each of the competing acts was pictured as if it supported the other.

Thus, functionalism did not try to attack the causes of the crisis of architecture, even less those of the crisis of the culture of the market. Functionalism was not concerned with whether the building was functional or not; that was the realm of mechanization of design. Functionalism was attentive to an *image* of reconciliation between mechanization and aesthetization. In this new conceptual framework, facts were supported by dogmatic base and backing, not of the archaic authoritarian type but of a modern kind: that of persuasion, suggestion, and illusion.

This was not the first time that the art of trompe-l'oeil had been applied to create the illusions of building materials. But certainly, the illusion itself was of a new kind.

Critique of Mechanization of Architecture and of Functionalism

Conceptual impasses and problems arise from the flux of human relations. Some of these problems are resolved, and some remain as persistent dilemmas until a new complex of human relations and new interests emerge. Functionalism, despite all creative efforts, has remained a half-baked solution, never overcoming the state of paradox and crisis from which it had been born. Slowly paradox and crisis became historical facts as architecture was confronted with new human figurations, confrontations, and alliances.

What we perceive at first glance as a battlefield of ideas, as a struggle of one norm against another, as a competition between conceptual systems–aesthetization against archaic design norms, mechanization against aesthetization–reflects antagonisms between social formations which, in their interest to dominate, devise ways of thinking in order to direct action in a particular way.

Mechanization of design develops concepts and techniques adapted to the behavior of natural objects and isolates the material aspects of the project from human relations. It was effective in controlling nature in a way unprecedented in archaic societies. But this conceptual revolution follows the lead of the market toward an atomized social order very different from the collective structure of tribes or even from the kinship or manor-dominated feudal society.

Today, we are not as confident as we once were about mechanization. We are ready to apply only partly such a limited way of thinking to design. But this does not mean that the grasp that mechanics has on nature should be loosened. It needs to be complemented, rather, by a more comprehensive approach.

A building project incorporates hours of work, a combination of human operations and efforts, but a project also acts on human ties and the flow and concentration of human power. Dependencies, domination, and reciprocity are all affected by the control a manmade object imposes on human behavior. These are issues that mechanization has not yet addressed. In addition, if there is anything mechanical about the mechanization of design–mechanical in the sense of rigid, naive, thoughtless, and unable to cope with complexity–it is in its failure to relate questions of human relations to the various elements of material reality. Buildings, their details and assemblages, have been linked with cost and function. They have been seen in terms of the behavior of the materials of their construction and performance. But from this viewpoint it appears only as through a glass darkly that each design product is a petrified human association and a human interaction to come. The reason for these new concerns arises from the twilight we may be witnessing of the market as the prime mover in society. In this respect, perhaps, we are in a situation diametrically opposed to that of the seventeenth century in Europe, with the market now retiring from the stage of the world theater.

Condorcet, the philosopher of the French Revolution, went beyond the confines of mechanics and aesthetics in design as early as the eighteenth century, when for the first time in the history of architecture he advocated the participation of the community of users in the decision making process in the design of a hospital.[48] The vision of Condorcet was to make the plan the outcome not only of the *Form* of Alberti, not only of the *Matter* of Galileo, nor of the *Arbitrary Beauties* of Perrault, but of an open dialogue in an assembly. This is the essence of humanistic design and its source of creativity.

New interests now require a more humanistic phase in the mechanization of design. Some of the existing conceptual tools will vanish. But as in the seventeenth century, which witnessed the dawn of mechanization of design, a large number of pieces of the new phase lie scattered around us. They await a new conceptual and institutional framework in which to gain identity, recognition, and currency.

NOTES

1. For the development of the history of engineering and of construction techniques, we suggest H. Straub, *A History of Civil Engineering* (1952); although not a recent book, it is still a good introduction to the topic. See also W. B. Parsons, *Engineers and Engineering in the Renaissance* (Cambridge, Mass.: 1968).

2. A. Gerli, "Riflesso per cangiar l'aria dei spedali nelle prigioni od in ogni altra publica coabitatione" in *Opuscoli* (Parma: 1785).

3. Cl. Perrault, *Recueil de Plusieurs Machines de Nouvelle Invention* (Paris, 1700, published posthumously).

4. Agricola, or Georg Bauer, *De Re Metallica* (Basel: 1556).

5. Domenico Fontana, *Della transportatione dell'obelisco Vaticano . . .* (Rome, 1590); C. d'Onofrio, *Gli obelisci di Roma* (1967); B. Dibner, *Moving the Obelisks* (Cambridge, Mass.: 1970).

6. L. B. Alberti, *De Re Aedificatoria* (1485), Book VI, *trans.* J. Leoni (Reprinted New York: 1966).

7. Hero of Alexandria, *Di herone Alessandrine de gli automati*, circa first century A.D. (published in Venice: 1589).

8. Le Corbusier, *Vers une Architecture* (Paris: 1923).

9. For further and more detailed reading on the application of this model, see A. Tzonis et al., *1650–1800: Les Systèmes Conceptuels de l'Architecture* (Cambridge: 1975), a research project funded by the C.O.R.D.A. of the Ministère du Culture, France. The work proposes a discourse analysis method for describing continuities, breaks, and innovations in the various levels of the conceptual systems of architecture in France in the seventeeth and eighteenth centuries. See also A. Tzonis, M. Freeman, R. Berwick, *Discourse Analysis and the Logic of Design*, GSD Publication Series (Cambridge, Mass.: 1978).

10. This basic division of the architectural history between "archaic" and "modern" pays more attention to the "meaning of architectural products, their use, the intentions of their creators and the attitudes towards them of the various social groups rather than the visual or technological organization." Looking at the factors that affected a major change in the evolution of thinking, we find that they are connected with the emergence of the market as the major determinant of human relations.

This view of modern architecture and of the process of modernization as a result of the commercial revolution differs fundamentally from the studies that hold the industrial revolution and the technological inventions as the moving forces in the development of architecture. Machinery and technological products are not an index of modernity in our approach. They are introduced "to increase the control . . . over the pace and flow of work." Keith Tribe, *Land, Labour and Economic Discourse* (London: 1978).

The basic periodization of history in terms of archaic and modern owes considerably to the work of Karl Polanyi on socioeconomic development. According to Polanyi there is a fundamental transformation of society and economy in history which has been carried out by the introduction of the market trade. This commercial revolution is the prime mover of human evolution. This model differs fundamentally from other approaches that consider the so-called industrial, urban, scientific, or demographic evolutions as the prime movers in history. K. Polyani, et al., *Trade and Market in the Early Empires* (Chicago: 1957). See in particular, "Aristotle Discovers the Economy" in K. Polyani, *The Great Transformation* (New York: 1944). See also G. Dalton, *Economic Development and Social Change* (New York: 1971).

11. For the Aristotelian ideas of mechanics see Aristotle, *Physics*, especially Book VII, and *Questiones Mechanicae*, attributed to Aristotle. For the history of the concepts of modern mechanics see P. Duhem, *Les Origines de la Statique*, 2 vols. (Paris: 1905–6), and E. J. Dijksterhuis, *The Mechanization of the World Picture* (Oxford: 1961), and M. Clagget, *The Science of Mechanics in the Middle Ages* (Madison: 1959). Alberti, in his *De Re Aedificatoria* (Florence: 1485), Book IV, Ch. III, discusses the size of the city in a manner analogous to Aristotle's: "Cities and ships should be by no means so big as to look empty, nor so little as to be crowded."

12. The treatment of the isoperimetral problems of Zenodoros can be found in the writings of Pappos. On the history of the problem, see G. Sarton, "The Tradition of Zenodoros," *Isis* 28 (1938), pp. 461–2, and in "The Years 'Forty-Three,'" *Isis* 34 (1942–43).

13. Aristotle, *Physics*.

14. W. Pauli, "The Influence of Archetypical Ideas on the Scientific Theories of Kepler" in C. G. Jung and W. Pauli, *The Interpretation of Nature and the Psyche* (London: 1955).

15. See "Letter to Hefulontius" quoted in Jung, *Nature and Psyche*. Additional references to the quarrel appear in Peter J. Amman, "The Musical Theory and Philosophy of Robert Fludd," *Journal of the Warburg and Courtauld Institutes* 30 (1967), pp. 198–227.

16. J. Kepler, *Apologia*, p. 460.

17. D. Phillips, "Paradigms and Incommensurability," in *Theory and Society*, 2, no. 1 (Spring 1975).

18. Kepler, *Apologia*.

19. R. Jaulin, *La Géomancie, Analyse Formelle* (Paris: 1964); E. Doutte, *Magie et Religion Dans L'Afrique du Nord* (Paris: 1909); A. Rouhier, *Traité Elementaire De Géomancie* (Paris: 1935). The archaic doctrines of African tribes reveals the same basic point of view; for examples see M. Griaule, *Dieu D'Eau* (Paris: 1948).

20. James Ackerman, "Ars sine Scientia nihil est, Gothic Theory of Architecture at the Cathedral of Milan," *Art Bulletin*, 31 (1949).

21. Nicole Oresme, "Le Livre du Ciel et du Monde," published in *Medieval Studies* (1941–43). Also *Tractatus de Latitudine Formarum*, by the same author, 2nd ed. (Padua: 1486) where we read: "En seront examinées plus clairement et plus facilement dès là que quelque chose qui leur est semblable est dessinée en une figure plane, et que cette chose rendue claire par un exemple visible est saisie rapidement et parfaitement par l'imagination des figures, aidant grandement à la conaissance des choses mêmes . . . . (Quoted by Duhem, *Les Origines*, p. 379). Oresme concluded that the graphic description of all problems can be reduced to six simple figures and their combinations. (Duhem, *Les Origines*, pp. 381–384.)

22. An English schoolman of the fourteenth century, Richard Swineshead, is ascribed the authorship of the *Liber Calculationum*.

23. Mersenne (1588–1648). R. Lenoble, *Mersenne ou la Naissance du Mécanisme* (Paris, 1943) is one of the most admirable studies of the transition between archaic thinking and the new science at the moment of "concordisme"–the efforts to bridge the doctrines of the Bible with modern rationality. Lenoble quotes fascinating problems of optimal "space allocation" applied to biblical discussions by Mersenne in his *La Verité des Sciences* such as "La géometrie des parallelogrammes nous fait comprendre comment les anges peuvent être à la fois au ciel et sur la terre . . . le calcul des surfaces, comment la vallée de Josaphat pourra facilement contenir tous les morts au jour du jugement dernier."

24. "Due lezioni all' academic Fiorentina circa la figura, sito e grandezza del Inferno di Dante" in J. Marchesini (ed.), *Le Opere di G. Galilei*, 9 (Florence: 1899).

25. F. Yates, *Theater of the World* (Chicago, 1969) contains references to John Dee and his *Preface*. Enrico de Anélis, *Il metodo geometrico nella filosofia del seicento* (1964) gives a very good survey of the impact of the Euclidean method on modern scientific thinking. See also N. M. Gilbert, *Renaissance Concepts of Method* (1960).

26. For an explanation of the technological developments related to social change see L. White, "What accelerated technical progress in the Western Middle Ages" in A. C. Crombie, *Scientific Change* (1963). For a point of view contrary to the one we present here, see A. Koyré, "Du Monde de l'Alea près à l'Univers de la Précision," in *Critique* (September: 1948).

27. E. Silbemer, *La Guerre Dans la Pensée Economique du XVIe Siècle* (1939).

28. Pico della Mirandola, *On the Dignity of Man*, trans. C. G. Wallis (1965).

29. The label neoplatonic was applied by historians who resisted the anachronistic interpretation of Renaissance architecture which saw Renaissance architectural theory of proportions and geometry as the result of theological preoccupations rather than of scientific or aesthetic thinking. It has become evident that the importance of theology in Renaissance neoplatonism has been exaggerated.

Most recently, attention has been paid to hermetism, due largely to the influence of such studies as Frances Yates's and Paolo Rossi's, as an alternative tradition in Renaissance architecture theory and practice.

Both views present problems. Neoplatonism and hermetism are tightly synchronized, which makes it difficult to isolate works which represent one or the other, and neither is capable of characterizing on its own the conceptual organization of the period, each being a partial incidence of the more global archaic conceptual framework.

30. The following statements by Galileo are quoted from the *Discorsi e dimonstratione matematiche intorno a due nuove scienze* (Leiden: 1638). We used here the translation by Henry Crew and Alfonso de Savio, *Dialogues Concerning Two New Sciences, Galileo Galilei* (New York: 1914).

31. Alberti, *De Re Aedificatoria* (Florence, 1485), Book I, chapter 1. This theme recurs in the writing of Plotinus who remarked that a house was first an idea and then a concrete reality in his *Ennead*, 1.6.1, quoted in E. Panofsky, *Idea, A Concept in Art Theory* (New York: 1968), p. 187. Thomas Aquinas in *Quodlibet*, IV.i.i., quoted in Panofsky, p. 196, and Meister Eckhardt in *M. Predigten*, No. 101, quoted

by Panofsky, p. 198, also shared the view that manmade objects were but the realizations of the same forms cogitated beforehand in the mind, and Ficino argued along the same lines that "the architect starts conceiving a notion ('ratio') of the building as an idea in his psyche." More extensive discussion of Ficino on this topic is in A. Chastel's, *Marsile Ficin Et L'Art.* It refers to Ficino's *Comment* in *Convivium* 5, and to Pico's *Commento,* II, 6.

One text is particularly explicit in presenting this archaic point of view in architecture, still alive during the Renaissance: ". . . one can increase the measures and numbers, but they should always remain in the same ratio. And whosoever should presume to transgress this rule would create a monster, he would break and violate natural laws." This is Francesco Giorgi's *Memorandum* for the Church of S. Francesco della Vigna, dated 1535 (quoted in Appendix 1, pp. 136–138 of R. Wittkover).

An extensive analysis of the concepts of Idea of Form is given in E. Panofsky, *Idea, A Concept in Art Theory* (New York: 1968), originally published in Berlin in 1924. There has been serious debate concerning whether or not Galileo ever actually transcended a belief in the doctrine of the supremacy of Idea over Matter. Thus his lines on geometry and nature from his *Te Saggiatore* (Rome: 1623) are often quoted to prove that he did not. The book of nature, he says, "is written in a mathematical language whose characters are triangles, circles, and other geometrical figures," and his reduction of the various building elements into prisms, cylinders, and cubes may recall the archaic, platonic and neoplatonic commitment to geometry, the framework that his interpretations of geometry fits into is new, reversing the archaic framework.

It is true also that Galileo refused to believe in Kepler's model of the elliptical orbits of the planets, preferring the Copernican circular model. This, several authors and especially A. Koyré in his *Etudes Galiléennes* (Paris: 1939) interpret as an evidence of Galileo's neoplatonism, his being a victim of "hantise de la circularité" (the haunting spell of circularity). Furthermore, Panofsky, in his *Galileo as a Critic of the Arts* (The Hague: 1954), refers to Galileo's correspondence with Monsignore Gianbattista Agucchi and to Agucchi's *Discorso del mezzo* as evidences of Galileo's attachment to "mezzanità," what we may call "middleness." However, the *Due Nuove Scienze* shows no trace of such tendencies. The geometrization of nature is only a means to justify not yet conceived perfect forms of the artificial.

32. Already in Aristotle's *Generation of Animals,* IV, trans. A. L. Peck (Cambridge, Mass.: 1942) monstrosity is still explained by reference to the domination of Form over Matter such as "whenever the Formal (*eîdos*) nature has not gained control over the Material nature." It is given an empirical, statistical definition: "the clan of the things contrary to Nature, although . . . not . . . in her entirety but only . . . in the generality of cases."

33. Legitimation is used here in the sense it is used in: J. Habermas, *Legitimation Crisis* (Boston: 1975). For him legitimation is "the process that elicits generalized motives–that is, creates mass loyalty but avoids participation."

The analogy between machines and architectural objects is to be found as early as in Vitruvius. He sees the design of both as determined only through proportions. He specifies the various parts of machines and of buildings using exclusively a module, without referring to size. In Book III, Chapter 1, on the design of temples he says "The composition of temples . . . depends on proportion . . . (on) a fixed module . . . for the parts of a building and for the whole." Similarly in Book X, Chapter 10, on the design of catapults he asserts that their construction depends on proportions and that all their dimensions should be determined "from the length of the arrow which the machine is to let fly." For example, the size of the opening of the frame is specified only as the ninth part of the arrow.

Luca Pacioli, a Renaissance mathematician and theoretician of design, has the same outlook. In the second chapter of his *De Divina Proportione* (Venice: 1509) the analogy between military machines and military architecture products is stressed and right proportion is seen as the common determinant in their design: "Se ben se guarda generalmente tutte sue artigliarie, prendise qual voglia, commo bastioni e altri repari bombarde, briccole, trabochi, mangani, hohonfee, baliste, catapulte, arieti, testudini, grelli gatti, con tutte altre innumerabili machine ingeni e instrumenti, sempre con forza de numeri mesura elor proportioni se trovaranno fabricati e formati. Che altro sono reche, torri revelini, muri antemuri, fossi, ponti, turrioni, merli, manteletti e altrefortezze nelle terri citta e castelli, che tutta geometria e proportioni con debiti livelli e archipendoli librati e asettati?"

34. Francesco di Giorgio, *Trattato,* Book IV (written between 1474 and 1482; published in Milan, 1967, under the curatorship of C. Maltese).

35. The complete history of military architecture has not yet been written. See Horst De la Croix, "The Literature of Fortifications in Renaissance Italy," *Technology and Culture,* IV, No. 1, Winter 1963; and from the same author, "Military architecture and the Radial City plan," *Art Bulletin* 4, 1962. Galileo himself lectured on military architecture and wrote two short treatises on the subject. See F. P. Fiore, "La città progressiva e it suo disegno," P. Marconi (ed.), *La città come forma simbolica, studi della teoria della architettura nell'Rinascimento.* See also M. Parent and J. Verroust, *Vauban* (Paris: 1971). The writings of Vauban have been edited by Colonel de Rocha d'Aiglun in *Vauban, Sa Famille, Ses Ecrits, Ses Oisivites, Sa Correspondence* (Paris: 1910).

36. There has been considerable and excellent research recently on this topic by Bruno Fortier et al. See, in particular, *Les Machines à Guérir* (Paris, 1976) and *Le Politique de l'Espace Parisien à la Fin de l'Ancien Régime* (Paris: 1975).

37. J. Bentham, *Panopticon* (London: 1787). See also E. Howard, *An Account of the Principle Lazarettoes . . .* (London: 1789), and R. Evans, "Bentham's *Panopticon,* an Incident in the Social History of Architecture," *Architectural Association Quarterly,* III, no. 2 (April–July 1972).

38. On the results of this unique empirical study, see in *Revue Générale De L'Architecture Et Des Travaux Publics,* X, 1852, p. 194, the *Rapport de l'Académie Royale d'Architecture sur la Provenance et al qualité des pierres employées dans les anciens édifices de Paris et de ses environs, demandé en l'année 1678 par Colbert, surintendant des Bâtiments* with introduction and notes by the Marquis de Laborde, Viollet–le–Duc, P. Michelot. See also Marquis de Laborde, *Mémoires et Dissertations,* 1852, and Ch. Sellier, *Rapport sur l'enquête ordonnée par Colbert en 1678, relativement à la qualité des pierres employées dans les constructions de Paris* (Commission du Vieux–Paris, 1898, p. 15–18). The belief that moonlight corrupted the stones of buildings goes back to an archaic astrological belief about the destructive power of the moon on the things of this world: See *Avicennae Liber Cannonis Medicinae,* Lib. IV, II, tract II, cap. II, referred to by Duhem in *Le Système du Monde,* II (Paris: 1914), p. 366. Even as late as 1702 in Paris, Michel de Frémin in his *Mémoires Critiques de l'Architecture* (p. 350) finds it necessary to title a chapter "If the moon eats the stones."

39. For a description of the project of the bridge see *Description des Projects de la Construction du Pont de Neuilly . . .* (Paris: 1788), by J. R. Perronet (1708–1794), the founder of the Ecole des Ponts et Chaussées. The construction of the church of Sainte Geneviève in Paris started in 1757. Gauthey and Rondelet belonged to the building supervisory committee applying exact scientific methods. After the French Revolution, the church was renamed the Panthéon. See E. M. Gauthey, *Mémoire sur l'Application de la Méchanique à la Construction des Voûtes et des Dômes . . . Relatif à l'Eglise de Ste. Geneviève* (Paris: 1774), and J. P. Rondelet (1743–1829), *Mémoire Historique sur le Dôme du Panthéon Francais* (Paris, 1797).

40. There is no major recent monograph on the observatory building. C. Wolf, *Histoire de l'Observatoire de Paris* (Paris: 1902) remains the best reference on the topic.

41. J. D. Cassini, *Mémoires Pour Servir à l'Histoire des Sciences et à Celle de l'Observatoire de Paris,* 1810. Cassini's text has been analyzed by Tzonis et al., *1650–1800: Les Systèmes Conceptuels de l'Architecture*

(Cambridge: 1975), p. 444. The argumentation in the text has been analyzed by A. Tzonis et al., *op. cit*, 328.

42. The best book on Claude Perrault is by W. Herrman: *The Theory of Claude Perrault* (London, 1973), although the author fails to explain in depth the cause for the antinomies in Perrault's thinking. See review by A. Tzonis, *Journal of the Society of Architectural Historians*, XXXV, No. 3 (October 1976).

43. B. F. Belidor, *La Science des Ingénieurs la Conduite des Travaux de Fortification et d'Architecture Civile* (Paris: 1739). A more modern edition with commentaries is by M. Navier (Paris: 1813). See in particular Livre I, "de la théorie de la maconnerie," p. 3.: "il faudra absoluement employer l'Algèbre and la Mecanique . . . (pour qu') on ne soit pas obligé d'attendre que l'expérience ou le hazard nous les apprenne."

44. Architecture is "the art of presenting images by the handling of forms," and it rejects science whose only purpose is to solve technical problems. For a full presentation of this point of view see E. L. Boullée, *Architecture, Essai sur l'Art* (1793–1799). The text (Bibl. Nat. Paris, MS 9143) has been edited by J. M. Pérouse de Montclos, Paris, 1968 and by Helen Rosenau (London: 1953). There are two interesting texts by C. F. Viel de Saint–Maux (1745–1819) that are symptomatic of the fully developed conflict in design: *Décadence de l'Architecture à la Fin du XVIIIe Siècle* (Paris: 1800), and *De l'Impuissance des Mathématiques pour Assurer la Solidité des Bâtiments* (Paris: 1805).

45. For the development and diffusion of technical knowledge see: F. B. Artz, *The Development of Technical Education in France* (Paris, 1966), R. Taton, *L'Enseignement et la Diffusion des Sciences en France au XVIII Siècle* (Paris, 1964) and J. P. Callot, *Histoire de l'Ecole Polytechnique*.

46. This is explicitly the subject of the first issue of the *Revue* of 1840. See section entitled, "Absence des conditions des progres de l'art de bâtir." Even today the opposition between mechanics and aesthetics has been perceived as a conflict between the "two cultures:" the culture of art versus the culture of science. Science is critical of the formal and decorative parts of buildings which cannot be explained through mechanics. Aesthetics condemn the rational skeleton of building: a prison that does not permit imagination to find a place. The theory of the "two cultures" is in fact only describing the conflict and gives no explanation of it. It does not investigate the causes of the conflict and obscures the fact that the two cultures have depended equally on the culture of the market since the Renaissance.

47. "Architecture is not building for the sake of building," said Frémin. The building is a "symbol" that "signifies that which has to be signified." The idea of apparent firmness in the building enters gradually into the academic architectural doctrine in the early eighteenth century without upsetting yet the presumption that architecture is a visual, formal exercise without reference to material and to size. J. F. Blondel, in his *Architecture Françoise*, I, p. 264, remarks in relation to Hôtel de Lambert ". . . mais il semble du moins que les trumeaux les plus larges devroient être placés aux extremités du bâtiment plutôt que dans le milieu, à fin que *la solidité parut en être l'objet*" [our emphasis]. In another instance he stresses that "Il faut que ce qui porte ait un *caractère* de solidité relatif à ce qui est au-dessus; que le délicat & l'orne ne soit pas sous le rustique & le simple . . . ," Ibid., I, p. 67. Thus the academic requirement that everything in a building must have an explicit justification is satisifed but the warranting norm is apparent firmness or "character" of solidity, both being visual, if not semantic, aspects of architecture.

In relation to the famous dispute on the correct use of erect architectonic elements, resting on horizontal or reclining planes, such as the balustrades of landings continuing up or down the staircase, Blondel takes again a functionalist stance, expressing in the adaptation of the element the requirements of apparent stability of the structure rather than finding a visual transformation of the balustrade, an "optical solution," the logic of visual consistency as Juan Caramuel did. "Pour . . . les balustres dont les moulures sont inclinés ou horizontales, nous dirons que l'exemple des bâtiments élévés sur une demi-côte ou sur une pente sensible, comme l'aile du nord à Versailles, & dont les plinthers & les corniches sont tenues horizontales, doivent servir d'autorité pour ne point admettre d'obliquité dans l'Architecture, sur-tout dans le cas dont il s'agit. Car ces balustres dont les moulures sont rampantes, offrent une idée contraire à la solidité que seule exige que leurs axe étant perpendiculaire pour porter le poids des tablettes, leurs moulures soient horizontales . . . ." For Caramuel de Lobkowitz see his *Architecture civil recta y obliqua* (1678). See W. Oechslin, "Oberservazoni su Guarino Guarini e Juan Caramuel de Lobkowitz" in *Guarino Guarini e l'internazionalita del barocco* (Torino: 1970), I, pp. 573–597. It is interesting to add here that the notion of "caractère" is used by J. F. Blondel in a sense very close to Frêmin's "signification:" "Toutes les différentes espèces de productions qui dépendent de l'Architecture devant porter l'empreinte de la destination particuliere de chaque édifice, tous doivent avoir un caractère qui détermine leur forme générale, & qui announce le bâtiment pour ce qu'il est." *Cours* II (Paris: 1971–1977), p. 229.

48. Condorcet, "Mémoire inédit sur les Hopitaux," 1786. Presentation by A. Tzonis, *Dix-Huitième Siècle*, 1977, 9, in *Le Sain et le Malsain*, ed. Jacques Guillerme. For a short history and critique of user participation and relevant techniques in architecture, see A. Tzonis and L. Lefaivre, "The Populist movement in architecture," *Dutch Forum*, 3, 1976, or *Bauwelt*, January 10, 1975.

1. Detail of the Parthenon on the Acropolis, Athens, Greece.

ARCHITECTURAL FORM AND THE TRADITION OF BUILDING

Cesar Pelli

Last December I went to Athens and the Acropolis for the first time. I had heard, read, and believed in so much of the praise given its buildings, especially the Parthenon, that I expected to be disappointed. I wasn't. Its beauty flooded my mind, my senses, and my deepest feelings.

I touched the carvings of a fragment sitting on the ground and felt their beauty in my fingers. I ritually kissed the marble that kept the delicacy of its forms for twenty-five centuries. These are sacred stones worthy of our continuous admiration.

The caryatids of the Erechtheion had been removed and replaced with cement casts. Visually the difference was insignificant, but the impact of the underlying reality was devastating. The building had been profaned by the imitation stone, imitation sculpture.

THE ARCHITECTURAL IDEAS that interest me most are those that help me to design better buildings. A coherent set of ideas—or theory—is necessary to do anything well. Indeed, it is impossible to take any action without a previous conception of the meaning of the action. We do not need to think about most things we do because we have accepted our community's standards and it would be impossible to rethink the reasons behind every action. In architecture it has been possible to do very good work, as in vernacular architecture, without having to rethink the ideas behind the forms, but in our times, when there is only polite architecture, in order to do any creative work we need to be aware of what is conditioning our forms and preferences. The clearer the conception, the better the architecture will be.

First we learn from our elders and then we learn from our buildings. As mature architects, nothing teaches us more than our own buildings. A building is an extraordinary laboratory for the development of forms and ideas. Architecture is such a complex art that intentions take form in our building that we did not know were there before. Sometimes our best ideas become apparent to us only after they are built.

In my development as an architect I have pursued several threads of exploration, but in this article I will focus on one critical concern of today: the relationship between a building's image and its reality and the underlying issue of legitimacy in architecture.

In our traditional past, buildings were made legitimate, that is, they were accorded a particular, high position that today we would call architecture, by their purpose. A cathedral was considered a major piece of architecture before it was designed. After it was built, it could be good or bad architecture, but it was always architecture. A building for storing grain or the house of a common person was never architecture, no matter what the quality of its design. Painting and sculpture were judged by very similar rules. All power, authority, and legitimacy came from God through princes and bishops and their organizations. This simple and reassuring condition started to be challenged with the Renaissance, and by the nineteenth century the process, and our social values, had changed. The legitimacy of a piece of art ceased to be given by the religious or social status of the object and started to require an intangible quality provided by the artist working within the rules of the art. The avant-garde represented the effort not only to change the rules but, more importantly, to have the artists themselves—with literary support—define them. In painting the shift of authority was complete. In architecture this was never quite so. Architecture continued having social demands and social responsibilities that the public at large and most architects never quite forgot.

Because the traditional hierarchical system had ceased to function, it was possible for the Modern Movement to develop a new ideology and a new set of rules to determine what is architecture and to give legitimacy to certain buildings and not to others. An almost universal act of faith was required to support these conceptions, and rules soon turned into dogma, but while it lasted there was no question about the legitimacy of buildings. It was relatively simple to know what was architecture and what was not and to differentiate between good and bad architecture. Today, of course, it is not so simple. The spell is broken.

The rules of the modern movement were brilliantly conceived and many of them were based on profound and still useful insights on the nature of architecture, but as a whole they could not cope with changing reality. They were too rigid, too utopian, and too centered on the beliefs of one particular time and place—western Europe in the early 1920s.

Today we are almost free of dogmas but we are left without a common yardstick for evaluating our buildings and without a solid base for architectural legitimacy. Legitimacy is essential, and its need can be perceived in the yearning of many architects to return to the forms of a past when there was no confusion and the hierarchical system was unquestioned. The hope is that by repeating their forms, somehow the value of the monuments of the past will rub off on the new building. I find the base for this attitude extremely weak, but as in all art movements, the weakness of the theoretical base does not preclude the possibility that some beautiful and brilliant buildings be built. My concern is that this cannot be a lasting condition or give a firm enough support for a new generation of architects to base their architecture on.

A healthy architecture does not look solid if it is void, does not look heavy if it is light, does not imitate inappropriate originals. Otherwise, after the momentary value of the fashion has passed, the disconnection with its underlying reality will vitiate the worth and legitimacy of any work so conceived.

What are we to do now? In my opinion it is too early for new all-encompassing answers. We cannot go from having discarded a questionable but brilliant and still useful set of dogmas to embrace flimsy new ones. We will not accept it. This gives us the opportunity to build from the ground up. The previous effort was an attempt at doing it all at once—godlike from the top down. Now our best hope is to build slowly, modestly, on the most solid ground we can find. And for me the solid ground, free of rhetoric, is given by the basic relationship between the art of our buildings and their system of construction.

We have a tradition of architectural design and a tradition of building construction. They are so interrelated that at times they have appeared to be the same. We know the tradition of architectural design. It is what touches our hearts and fills our history books. The tradition of building refers to the process and technology of their construction, the materials we use and how we use them, and the laws and conventions that set the parameters for their use. This tradition changes at a much slower pace than the traditions of architectural design that grow from it. In the Western world the dominant building tradition had remained basically the same since ancient Egypt until the nineteenth century. It is the tradition of building in sculpted stone masonry. Throughout its history stone was a special, almost sacred material, each block carefully chosen and lovingly carved. By dominant I mean the tradition of construction of the important buildings of a society, the buildings that set the tone and introduce changes in style that are imitated in lesser materials. This tradition of construction was at the root of several different Western architectures, setting their limits and defining their opportunities and, because of this, establishing a strong kinship between them. I am referring to Egyptian, Greek, Roman, Romanesque, Gothic, Renaissance and post-Renaissance architectures, but one could also include the sculpted stone masonry architectures of other places and times.

This tradition started to be challenged by newer and radically different techniques in the nineteenth century, but it remained alive and competitive until the second world war. Since the war and because of the great social, economic, and technological changes it produced, that tradition has ceased to be. It is not that stone buildings are not possible, that stone is not used and that some sculpted stone masonry buildings are not being built. It is that it has ceased to be a living tradition animating our most important pieces of architecture and being informed by and responding to the minds and demands of the best architects of our time.

Today a new tradition of construction has taken its place and has become the dominant tradition. This new tradition is of a very different nature. Whereas the old tradition of stone construction consisted of building up with thick and heavy horizontal courses of stone, the new tradition of construction consists of erecting a frame of steel or reinforced concrete to support a lightweight envelope that encloses the building. This envelope is not only lightweight but is made of several thin layers, like layers of clothing. The layers are quite specialized in their purposes: physical protection, vapor barrier, thermal insulation, waterproofing, interior finish, exterior finish plus others that may separate or hold up these materials. Because the wall does not support the building, its plan has no poché, and it gives no clue as to the height of the walls. The purpose and role of the exterior wall has also changed. Buildings today are not just shelters but are complete containers of controlled environments requiring continuity of the enclosing envelope.

Logic and common sense tell us that the aesthetic systems of architecture that grew from a tradition of construction based on sculpted stone masonry must be, in their basic nature, radically different from the systems of forms or architecture that can grow from a tradition of construction based on this more specialized assemblage of a skeleton structure with a multilayered, thin, enveloping enclosure.

One may say, "How about the Greek temple and its forms, derived from an earlier tradition of wood construction?" Because this is a hallowed precedent, it would require a long argument to satisfactorily prove that the example does not apply. I will touch here only on some key issues.

First, the characteristics of heavy timber construction for short-span buildings are not dissimilar from post and lintel stone construction. This is why the primitive hut of the Abbé Laugier is almost a Greek temple. Second, the building type and its forms had an uninterrupted history supported by and supporting a living religion. Third, the architectural forms of the Greek temple are fully consistent with its stone nature; only the formal elements that were suitable to express an architecture of stone were kept from a previous tradition. When we are in their presence it is only the stone that matters; the wood precedent becomes only a historical curiosity with no architectural reality.

We face a new condition. First, the present tradition is radically different in its nature from the old one. It is, however, akin to several vernacular and local traditions such as the New England frame house—or church—sheathed in clapboard or shingles. Second, the development of our building types not only has been interrupted, it has been transformed beyond recognition, and the chain of architectural forms that grew from our living religion was broken and abandoned by the Renaissance revolution.

Architecture can grow only on a living tradition of building. There are several regional living traditions today, but sculpted stone masonry, historically the most important and universal, has ceased to be, and the dominant universal tradition that has taken its place is so new that its nature is still not well understood. This condition can be disconcerting, but it is also full of opportunities. Want it or not, we have to participate in developing the systems of forms for a still very young architecture. In this endeavor we have to reconnect with our modern past, including all the good efforts of the nineteenth century, but we must do it selectively and with open eyes. We can also reconnect with our older history, but we should avoid the formal elements that are specifically descriptive of a reality that no longer exists. The building tradition is given to us; how we use it to give life and soul to our architecture is our responsibility.

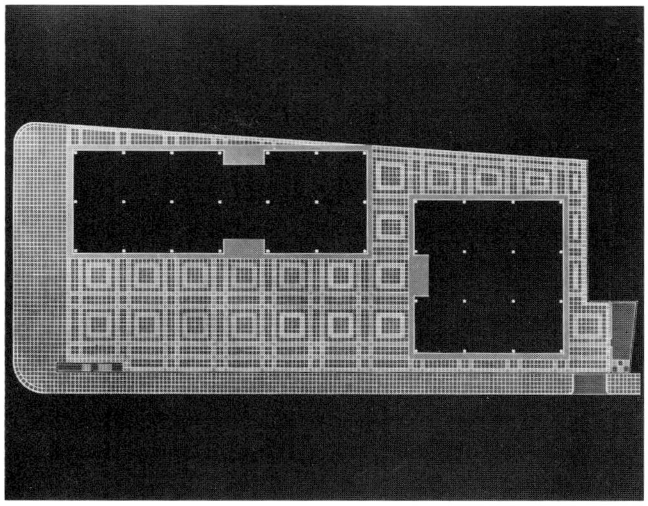

2. Plan, Century City Medical Plaza, Century City, Los Angeles, California (designed with A. J. Lumsden while at the firm of Daniel, Mann, Johnson and Mendenhall).

The tradition of building does not, by itself, produce architecture. It only sets the limits and opportunities for it. Within those boundaries there exist infinite possibilities. It takes the intuition of an artist to bring them into being.

As I look back at the buildings I have designed, I can see now that I was not only trying to design the most intelligent, responsible and beautiful buildings I could according to their circumstances and my talent, but also that there is a continuous thread of a concern about the relationship between my architecture and the building tradition of our times. I will now follow this thread as it moved through several of my projects. It is in the process of bringing our buildings to reality, where we architects can learn something about the nature of what we do.

Century City Medical Plaza, Los Angeles

One of the first of my designs that was built was this medical office building in Los Angeles (Fig. 2). I designed it with Anthony Lumsden in 1966 while I was at DMJM. It was a very limited problem with a minimum budget. The plan is a structural nine square with the center square occupied by the vertical core. I believe this is the first high-rise building completely enveloped in a skin of glass from the ground to the sky.

The design intention was to make architecture by clearly expressing the role of the skin as enclosing a volume of controlled space. To achieve this purpose we developed then a number of design elements that together have become the norm for this type of building and have been repeated thousands of times in bad and good structures. These are some of the elements

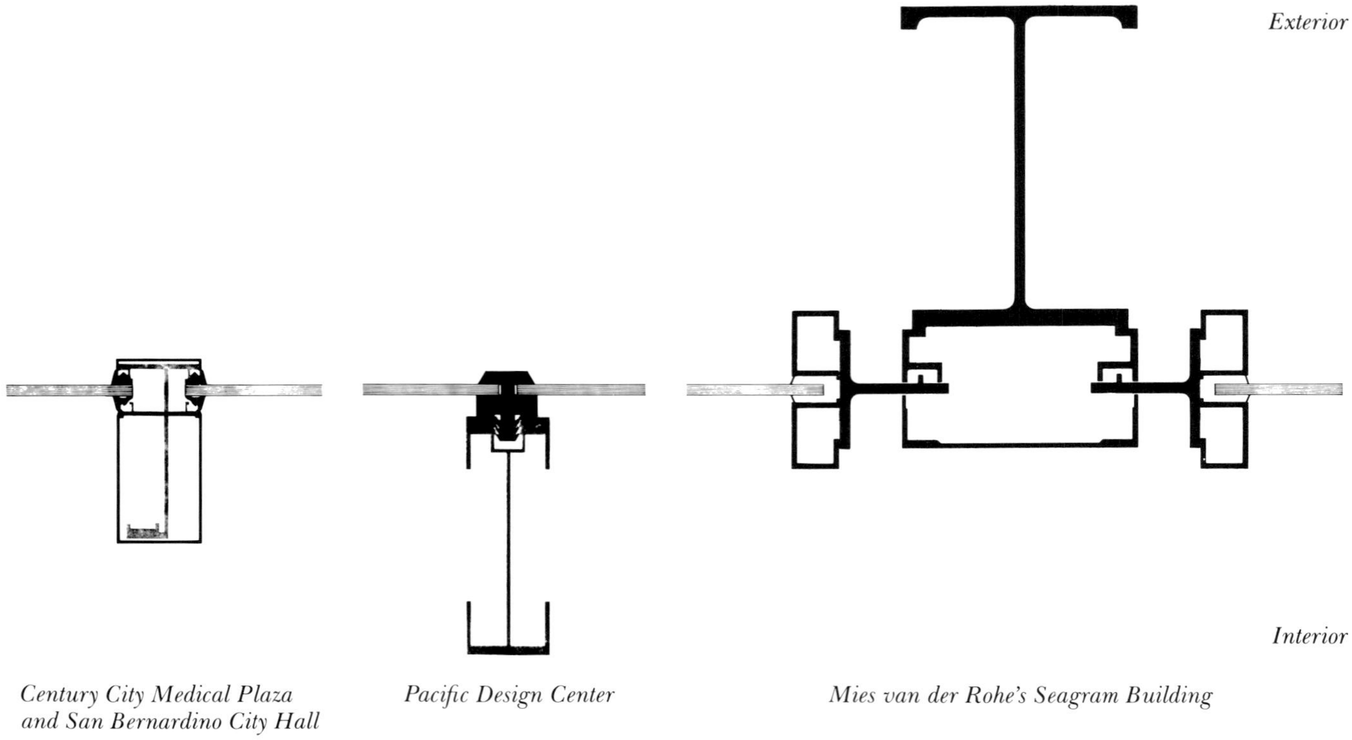

3. Mullion details.

we developed: Minimum projection mullions (they allow the surface of the glass to continue past them and establish the preeminence of the skin over the armature—see Figure 3, comparing one of these mullions with one by Mies van der Rohe); a uniform grid of mullions throughout the exterior (it expresses the character of the space enclosed as undifferentiated, subdividable, and modular, and it visually describes a two-dimensional exterior wall enclosing a three-dimensional volume); ceramic glass spandrels, colored to match the vision glass (it unifies the wall while allowing for straightforward structural and mechanical systems); minimum corner detail (the corner mullion is of the same magnitude as any other, denoting that this is a single enveloping surface folded around the edges of the volume and not four facades) (Fig. 4). These are rather simple elements. Many had been around for some time; others we developed first for this building. When they are used together they make a very clear architectural statement. It is a most logical and efficient way of enclosing an office building today and because of this it has been repeated with variations in almost every city of our country and in many foreign ones.

This project was primarily experimental: how to make something out of almost nothing and how to make an architecture that directly addressed the nature of our buildings. As an experiment it was very successful, but it is not a good piece of architecture. The economic and practical limits were too severe to allow it, but I have questions about the architectural intentions as well. The design is most clear about its building nature and the abstractness of the gridded wall is aesthetically and intellectually satisfying. It is that very abstractness, unrelieved by the simple prismatic form, that removes the building not only from its surroundings, but also from direct contact with our senses and our emotional expectations. When this design is repeated in mirror glass (as it often is), the building removes itself even further, becoming an alien presence.

COMSAT Laboratories, Clarksburg, Maryland

I designed this building in 1967 while still at DMJM. The complex and differentiated functions of the laboratories are organized along a public spine that allows each of the constituting elements to take a form suitable for its need, and to grow and change independently of the rest of the building (Figs. 5–7).

The laboratories are enclosed in a skin of aluminum and glass. The windows of the private offices are "punched holes" in the aluminum panels. The windows in the public or common spaces form continuous bands. The aluminum skin turns over, under and around in the three dimensional planes, suggesting a continuous wrapping of the complex volume and making visible its role independent from the structure.

I arrived at the decision to design this modular skin of aluminum and glass not only because of its intellectual and aesthetic appeal but also because of the suitability of its character to the purpose of the building: to be a place where research, experimentation, and construction of communication satellites takes place. The problem was made simpler by the fact that

4. *Century City Medical Plaza.*

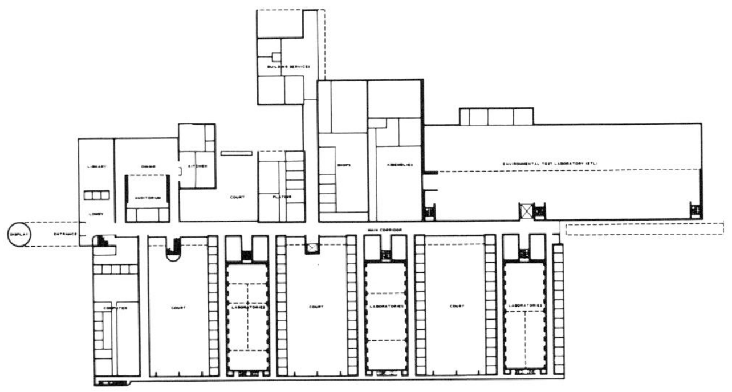

5. *First Floor Plan, COMSAT Laboratory, Clarksburg, Maryland.*

6. *Exterior view, COMSAT Laboratory.*

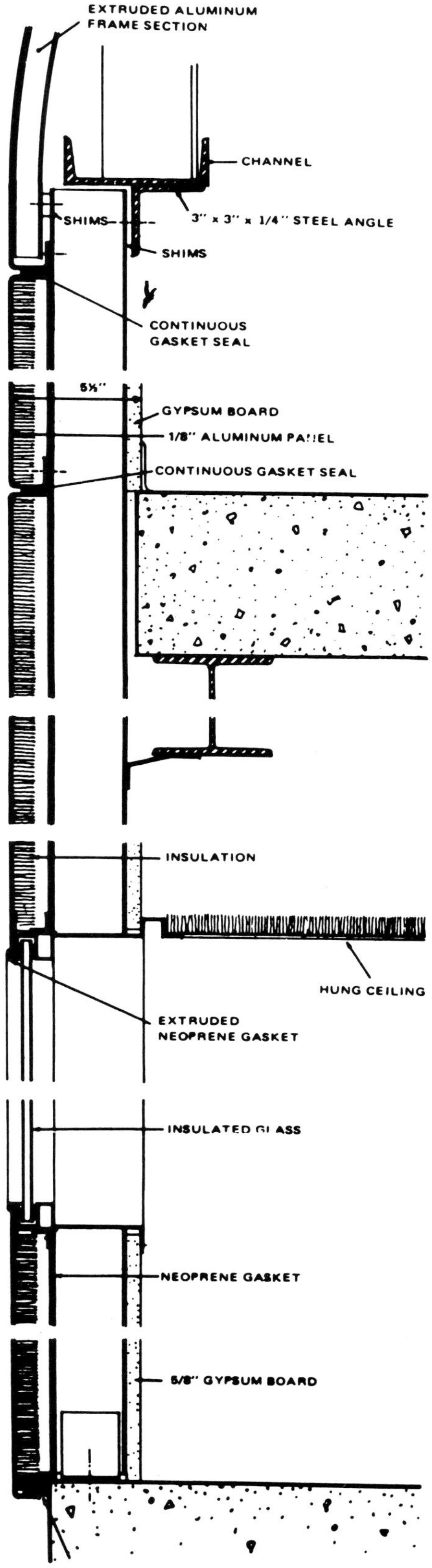

7. *Typical Wall Section, COMSAT Laboratory.*

8. *San Bernardino City Hall, San Bernardino, California.*

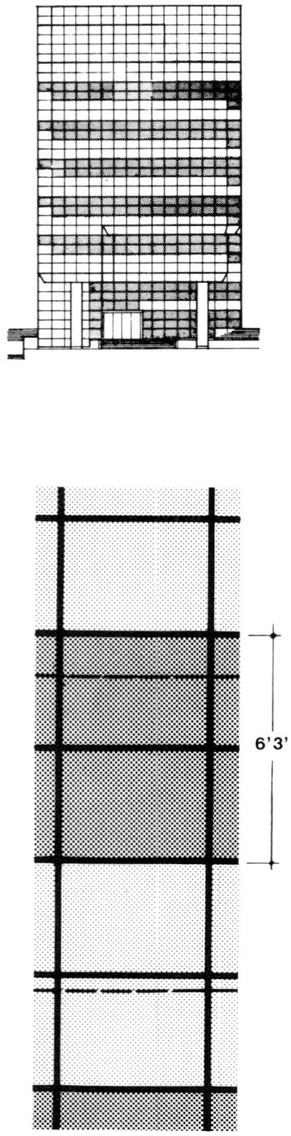

10. *East elevation, San Bernardino City Hall.*

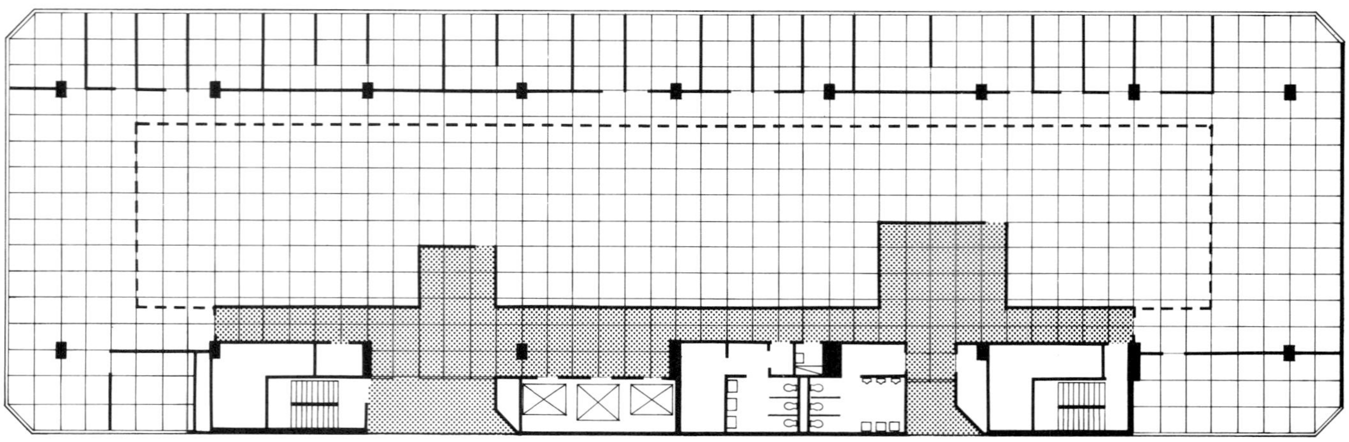

9. *Plan, San Bernardino City Hall.*

the building was built on a large piece of ground of nearly 200 acres with no visible neighbors. In this building the wall is made of two materials, but both are light and thin and they are visibly so. They are also industrially produced materials, only recently available in these sizes and for these purposes. The wall is obviously a product of our times, like the satellites, and like them the building celebrates a technological achievement. Today, when a great diversity of materials and systems of exterior wall construction is available to us, the matter of their appropriate character for the purpose of the building and for the place where it is built becomes a critical issue.

City Hall for San Bernardino, California

A few years later (1969), while at Gruen Associates, I designed this City Hall (Figs. 8–10). It was a good problem, giving me greater opportunities than the Medical Center. The spaces were not undifferentiated as in an office building and we had to provide for public functions and ceremonies. The plan is linearly organized in the east-west direction with all the support elements on the south wall. This achieved two purposes: first, to give all habitable spaces views of the mountain range north of the city; and second, to avoid placing them facing south, exposed to the fierce sun of San Bernardino. The building is at a pedestrian crossroads and it gathers these movements under itself as though in a monumental portico directing them to the main entrance.

The San Bernardino City Hall is not a box but a complex form engaged in its surroundings and fulfilling most of the expectations a visitor may have. In this building the architecture is not as abstract as in the Medical Center, but the intentions are just as clear.

A regular grid defines the enclosure of the interior volume. The grid is smaller than in the Medical Center to allow for changes in the amount of vision glass in different orientations and for different functions. The small grid also allows for an intermediate 45° plane between sides of the building, making the continuity and singularity of the surface envelope unmistakably clear. The glassy grid continues on the underside of the portico as well, reinforcing the surface continuity and fully expressing the nature of the building as a tightly sealed envelope containing a volume of usable and controlled environment.

Two rows of balconies on the south face of the building produce "punctures" in the skin, playing against its tightness. They also read as if there were a second glass form inside the main one. Elements like the balconies and portico give the building some scale. The City Hall is also perceptually very rich and it involves the visitor in changing relationships as one moves into, towards, or through it.

I have one concern about the design approach of the City Hall. To achieve visual uniformity between opaque spandrel panel and vision glass window one needs to use either tinted glass and matching spandrel, which result in a building that reads almost black, or a reflective coated glass for windows and spandrels that read light in color but are even more abstract in quality. Because of this, I explored other possibilities in my later designs.

United States Embassy Office Building, Tokyo, Japan

I designed this building in 1972 while at Gruen Associates (Figs. 11,12). The nature of its function, its place in Tokyo and the requirements of our client, the Office of Foreign Buildings of the U.S. State Department, led me into some useful explorations and to a visually simple but intellectually complex design.

Seismic conditions mandated that the wall be separate from the structure and supported by it. It was also required that the exterior wall be made of concrete and that all windows be operable. There was no conflict because the nature of a curtain wall or building envelope is not defined by the materials it is made of but by its role as an enclosure that does not support the building and is itself supported by an internal structure.

If we want to express the exterior wall as a lightweight continuous envelope, nothing does it better than glass. An all-glass wall is not only thin, light, and fragile, but it is visibly so. The image is one with the purpose. Making the solid panel of the same material as the window is also the only way to achieve a single continuous surface, leaving only the form of the volume to express.

In this building there are two types of opaque panels: one is flat and modular, interchangeable with the operable windows, and is applied to the exterior of the structure like a curtain wall; the other is modeled with horizontal ridges and fits between the structural elements of the end wall. The building is a balanced composition of large wall surfaces with a heavy three-dimensional structural grid. The grid acts as entrance portico, and it helps to define the character of the curtain wall in contrast to itself. The wall is composed in horizontal bands of windows and spandrels interrupted by the vertical core of elevators.

The design approach for the U.S. Embassy Office Building in Tokyo is therefore less obvious, less clear, and less didactic than the one for the City Hall or for the Medical Plaza, but it is richer, more responsive, and capable of a greater expressive range. The building expresses its limited ceremonial purpose very well. It is not the Embassy of the United States; that is where the ambassador lives. It is only the Embassy Office Building. The building had to be an American building, but one that would be sympathetic to the country and site where it was built.

I am also very pleased with the character and expression of the concrete, operable-window curtain

11. U.S. Embassy Office Building, Tokyo, Japan.

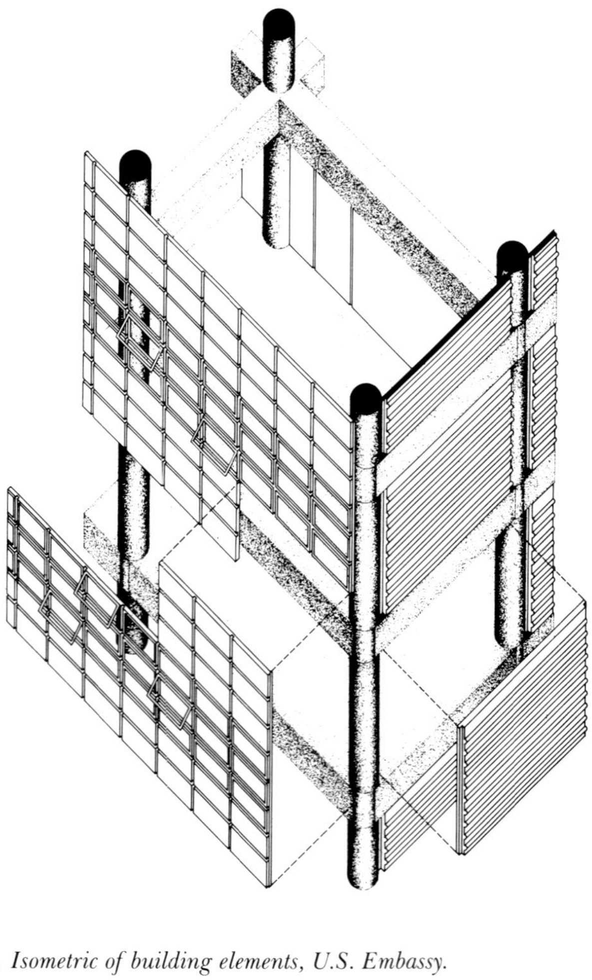

12. Isometric of building elements, U.S. Embassy.

wall. It is one of the most taut and crisp walls I have designed. Its relationship to the structure is made into architecture at the grand entrance colonnade and is also expressed at the ends where the system is interrupted as it corresponds to a thin slab derived from a plan organized on a line of circulation.

Pacific Design Center, Los Angeles

The Pacific Design Center or "Blue Whale" is probably the best known of my buildings (Figs. 13–15). I designed it in 1972 while at Gruen Associates. One decision in this building had the greatest design effect: to sheath the building in blue ceramic glass.

The effect was not only to make the building noticeable but also to shift the aesthetic content of the enclosing skin. In the Medical Plaza and in the San Bernardino City Hall, the gridded gray or brown walls provoke an aesthetic pleasure that is primarily intellectual: the cartesian grid as the graphic expression of human reason takes tangible form and defines a volume of habitation at monumental scale. The perceptual richness of the reflections on the glass is not sufficient to diminish the clarity of the main reading. The intense color does it. We see the Pacific Design Center first as a great blue silhouette. Only later do we become engaged in other issues. That is, we see it first with our senses, then with our minds.

This choice was, of course, very purposeful. What is important to note is that this blue enclosure is as logical as that of any of my previous buildings. There is no mistaking that this is a building of our times, with an image that is fully consistent with and supportive of its rather new function (a national showcase for the interior design trades) and with the nature of its construction. I believe it is also a positive answer to its site: a large empty parcel in a nondescript neighborhood going through a process of rapid change and needing a signpost or landmark to let us know where we are when we are there.

The Museum of Modern Art: Renovation and Expansion and Museum Tower, New York, New York

This was the first commision I had after I opened my own office in New Haven, Connecticut, in January, 1977 (Figs. 16–19). This was, and is, a difficult and rewarding project touching on many architectural issues of importance today, but for the purposes of this article I will concentrate on one–the building enclosure.

Our charge was to produce a design that would harmonize with the existing buildings by Philip Goodwin and Edward Durell Stone (1938) and by Philip Johnson (1958 and 1964). The residential tower was to relate in design to the museum buildings so that the total would look like one composed whole but with two distinct elements. The design was to fit comfortably in its place in midtown Manhattan. For obvious

13. *Pacific Design Center, Los Angeles, California.*

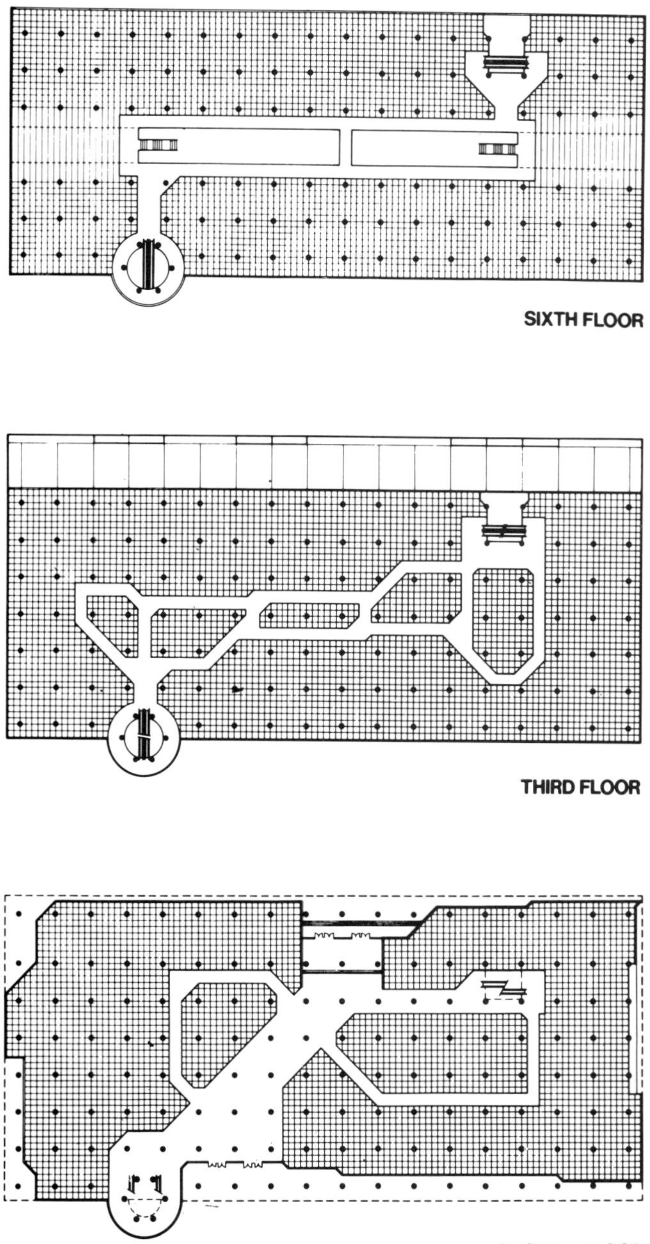

14. *Plans, Pacific Design Center.*

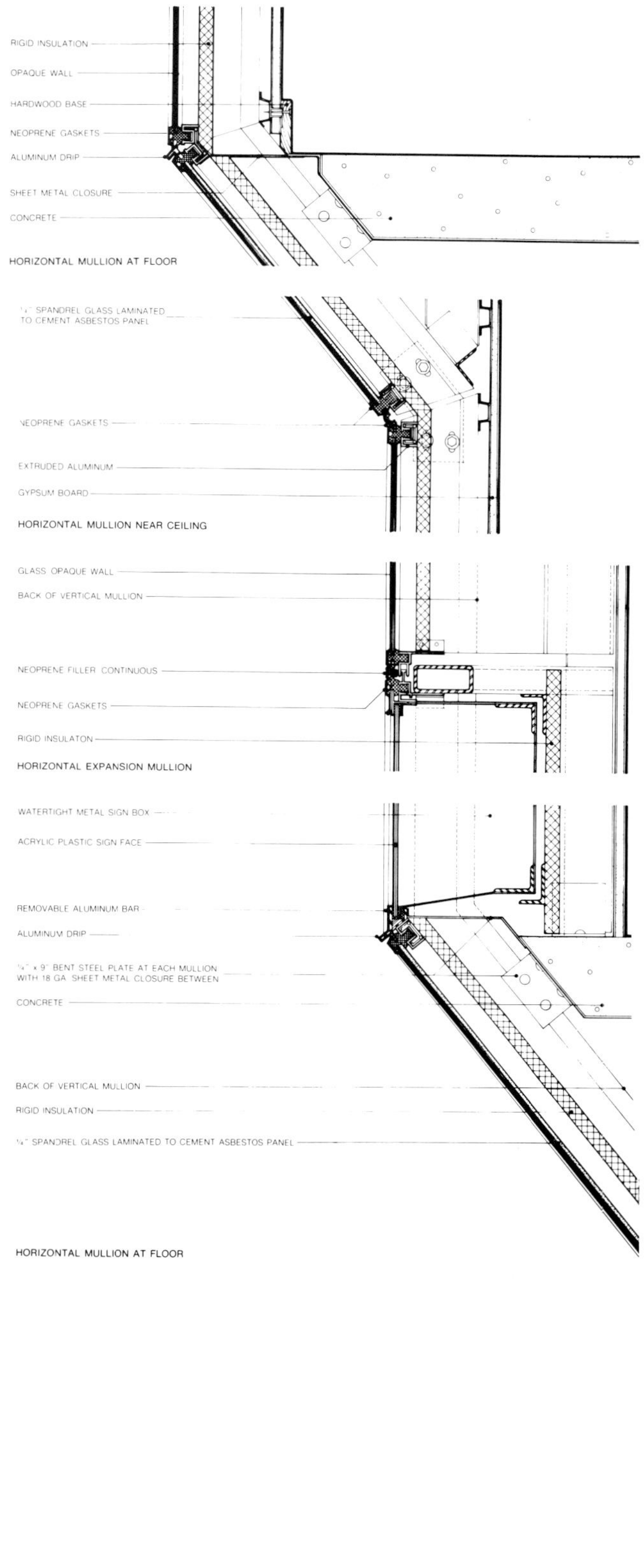

15. *Horizontal Mullion Details, Pacific Design Center.*

16. Model, Museum of Modern Art, New York City, New York.

17. Detail of model, Museum of Modern Art.

reasons the design had to reflect what The Museum of Modern Art stands for in the world of art.

The buildings are now almost finished. They are enclosed in a wall of vision and ceramic glass. The vision glass is tinted as is necessary for apartment windows and to continue the tinted glass windows of the East Wing and No. 11 buildings. The ceramic glass has been produced in eleven shades of color to match the samples we provided. The colors are predominantly brownish-grays to fit within the color range of this area of Manhattan. (Now they are up and they do fit in very well.) The different colors are used not only to produce a beautiful wall (it is) but also for other reasons, dearer to me. The colors of the wall of the new museum wing on 53d Street are chosen to balance the black/brown wall of the East Wing, leaving the No. 11 building to play its intended 1938 role of introducing modernity to Manhattan. It continues to be a white billboard in a street of dark brown buildings, the main element in the composition, and it continues to be the public entrance to the museum.

The different colors allowed me to express the different functions in the buildings and some of their character. The new museum walls have a large institutional scale, extending the alignments and rhythm of the No. 11 building. The tower expresses each floor and also the position and scale of the apartments and the rooms within them. There is no question that this is a residential tower.

This building represents for me a greater shift of the primary intention of the wall than did the Pacific Design Center. The MoMA buildings introduced in my architecture representational objectives. That is, the design responds also to our memory and preconceptions, not just to our minds and our senses. However, the building is still enclosed in a clear and logical curtain wall not pretending to be anything else. It is taut, light, thin, crisp, and quite beautiful as a polychromed, polished pylon.

The Four-Leaf Towers in Houston are also good examples of the same approach, although with some important secondary differences: First, the color code is constant—dark red-brown for columns, very light gray-brown for floors and window sills, and medium red-brown for opaque wall panels. Second, the mullions change color to strengthen the color fields of the wall. Third, the envelope continues at the top to complete the form against the sky. Fourth, all four façades are identical and symmetrical; this feature, plus the tops, makes the buildings into pylons (Fig. 20).

World Financial Center, New York, New York

Since the time I was designing the U.S. Embassy in Tokyo, I have been interested in the possibility of doing a curtain wall of stone. The interest is obvious: stone is the material that characterized the cherished buildings of a past that is no more. It could be tempting to design a building today clad in stone and im-

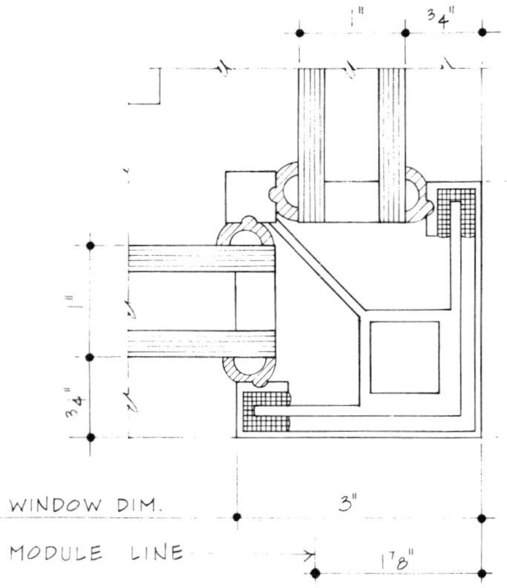

Typical Corner Mullion

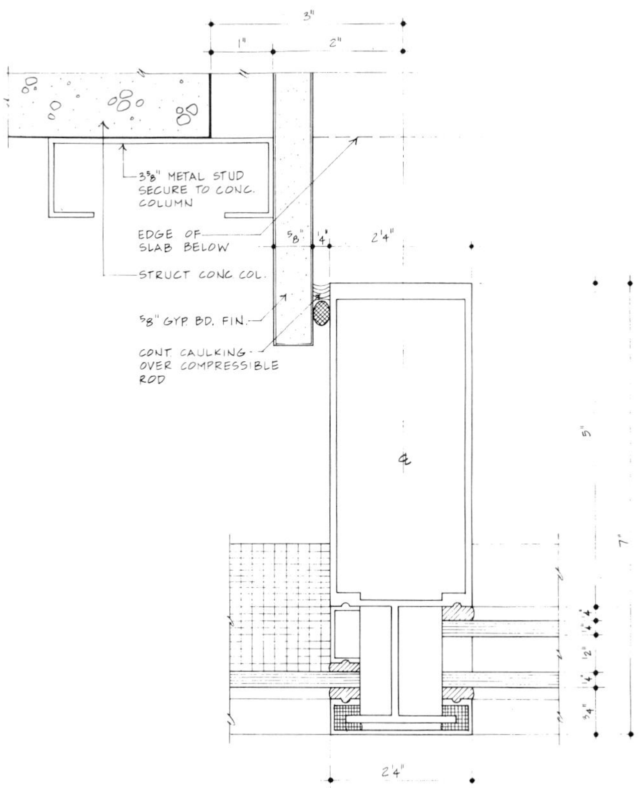

Typical Mullion, Non-Vision and Vision Glass.

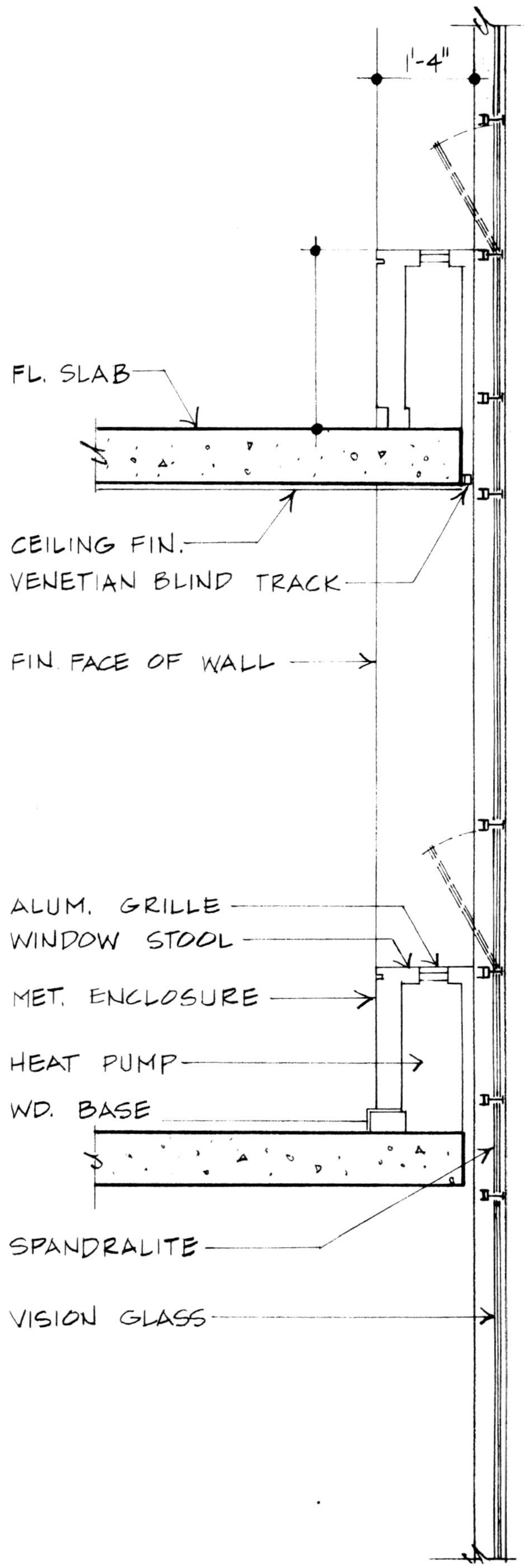

18. *Mullion Details, Museum of Modern Art.*

19. *Wall Section, Museum of Modern Art.*

20. *Four-Leaf Towers, Houston, Texas.*

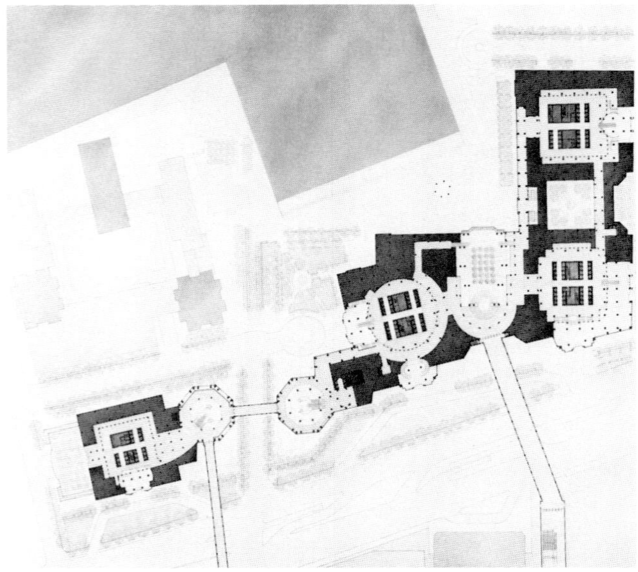

22. *Plan, World Financial Center.*

21. *Model, World Financial Center of Battery Park City, New York City, New York.*

itating the forms of the past, but today we use stone in such a different manner that its basic nature as a construction material has changed and the change is radical. To use it well requires a good understanding of the differences. The exciting challenge therefore is to use it with some rigor, respecting its new nature and still producing architecture. These are some of the key differences: Stone used to be the structure; now it is supported by a structure of steel or concrete. Stone used to be cut by hand to fit; now it is fabricated by machines in a shop far from the building site. Stone was used in thick blocks; now because it needs to be transported long distances and then hung high up on a structure, it is as thin as possible to save weight. Stone used to be carved and sculpted; today it comes basically in thin slabs like plywood; anything else is extraordinarily expensive. Stone used to be the whole wall; now it is only the exterior layer. In the past there were buildings of stone; today there are buildings of steel and concrete with a stone veneer.

The new building tradition is not a lesser or a greater one; it is just different, with its own opportunities to be developed. The problem of designing a building clad in stone appeals to me because it is at

23. Entrance Rendering, World Financial Center.

the cutting edge, where mistakes are easy and the rewards are great.

I have had a few commissions in recent years where I could try to make architecture using stone as cladding. The World Financial Center and the Clinic Building for the Cleveland Clinic Foundation are both good examples and both are under construction. The design for the World Financial Center had to respond to more demands and it addresses more issues; therefore, it is a more interesting example to review.

The buildings are designed to fit in and complement the city form of downtown Manhattan and to relate comfortably to the World Trade Center Towers and to the river edge. The aesthetic relationships are established at two levels: one responds to compositional and geometric characteristics and the other responds to the emotional and symbolic qualities of that part of Manhattan (Fig. 21).

These are some of the compositional/geometric responses: the buildings are aligned with the Manhattan street grid to the north, rotating towards the south to align with Westway and the river edge (Fig. 22). The buildings have marked setbacks at the third, ninth, and twenty-fourth floors to correspond to the prevalent heights and setbacks of the buildings of the area. The dimensions of the shaft of the tower are approximately those of the towers of the World Trade Center. These are important relationships that give form to the urban aggregation. We become keenly aware of their need when a design disregards them, for example, by positioning the building on a diagonal alignment, and disrupting the city order for blocks around it. However, these compositional/geometric responses are not enough. We all know buildings that have been carefully composed to match basic and subtle alignments but that still remain alien presences. To make the building of its place, an additional and different type of relationship is required. In our design for the World Financial Center the towers have a strong vertical thrust and they culminate in copper-clad roof forms, or crowns, that give them distinctive silhouettes. They represent an attitude of optimism and celebration which is what we have always expected of skyscrapers. This attitude is perhaps the key element in making the World Financial Center towers part of downtown Manhattan, but there are others that will contribute greatly: the quality of the

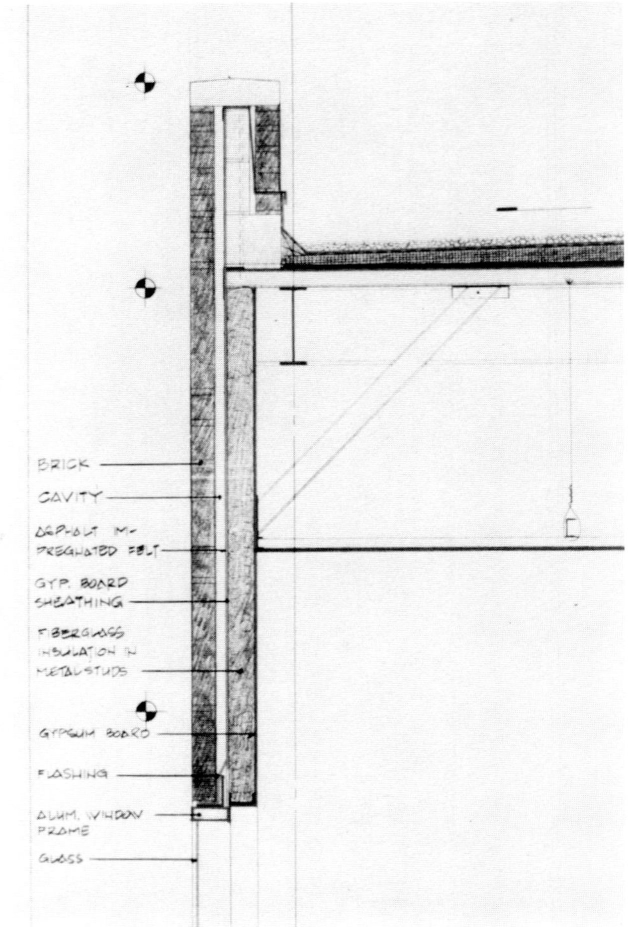

24. Wall Section, Herring Hall, The Jesse H. Jones Graduate School of Business Administration, Rice University, Houston, Texas.

surface material (flame-finished granite) and its color (warm gray); the colonnaded base; the large formal entrances; and the use of glass separated in discrete window units (Fig. 23).

The buildings are thus connected to their place. They are also connected to their time and to their own reality, which is at least as important. The wall is glass and stone, but at every critical instance its quality as a lightweight envelope is revealed. The design never uses elements that would confuse this reading or pretend that these are built as they are not. The buildings are further connected with their time by acknowledging that we have learned much about the nature of these vertical places of work in our recent past. The grid of stone expresses the modular nature of the undifferentiated office space, the façades are equivalent, and the central tower is a perfectly square prism. The crowns are enclosures of mechanical equipment, expressing thus that they are not of the same nature as the office space. The formal entrances do not mark their role with a gesture derived from opening a hole in a mass of stone such as an arch, lintel, pediment, or keystone; instead, they celebrate the transition from outdoors to indoors by appearing to be the peeling away of several layers. In this instance, as in many others, it has been my goal to achieve experiences as rich and complex with our building tradition as those afforded by the architectures that were based on a tradition of sculpted bearing stone. The intention, now as then, is to work within the basic nature of the tradition and not to express every construction characteristic of each building.

Graduate School of Business Administration, Rice University, Houston, Texas

This is the most recent of my designs and the one in which the ideas I have been expressing were put to their most severe test. Rice University has one of the most handsome campuses in this country. Its charm comes primarily from the Master Plan and the character of the buildings designed by Ralph Adams Cram at the beginning of the century. The best of the later buildings have been respectful of his original intentions.

We were engaged as architects with the condition that we would design a building that would be compatible with the older ones and supportive of the environment they create. I fully agreed with this intention to make the new building, Herring Hall, of its place, but I added that I would attempt to do so while making it also a building of its time.

We sited the building so as to best continue the Cram Master Plan intentions, and we organized it as bars slipping past each other as in the earlier buildings. As per our brief, the roof is of clay tile and the walls are of beige "Rice brick" in a Flemish bond pattern with repeating shiner courses and limestone base and details.

We quickly discovered that the brick wall had to be a four-inch veneer supported on steel studs to bring the building within its budget (we did). Although not technically so, I refer to it as a brick curtain wall to make its character clear. Because the wall is a veneer there were many architectural forms associated with brick that I could not use. What I could do was to introduce patterns in the wall through the shiner courses, limestone, and color tile accents and patterned areas of a reddish semi-glazed brick (Figs. 24, 25). These patterns and other simple devices allowed me to develop a rather rich vocabulary with which Herring Hall could engage in friendly conversation with the older buildings on campus.

Herring Hall is clad in traditional materials, but they are used in contemporary ways. The structure is a steel frame 14 feet on center, spanning the volumes. The brick wall is always expressed as a nonbearing skin with windows flush with the wall and on a 14-foot module. The end walls are stops and fillers in an otherwise continuous system, and they are detailed to be read as such. The decorative elements used are those that conform to the process of construction of this building. This process is not the dominant tradition I have been describing, but is a related regional variant of a conventional, postvernacular tradition suitable for smaller, lower buildings such as this.

25. Typical Bay Elevation, Herring Hall.

• • • • •

In my designs my interests have shifted in focus from the San Bernardino City Hall to Herring Hall at Rice University. The designs are not only addressing new issues but they are including more of them. They grow in a different direction, but they grow from the same root and all of them remain as viable possibilities for my architecture. Taken together, they are starting to make a well-fronded tree.

I have followed my own thoughts and intuitions from building to building but I have also been participating in the architectural discourse of my times, affecting it and being affected by it. This article is further participation and an attempt to influence the direction of this discourse.

I have tried to be as objective as I could about my design explorations to make them useful to others. The design of a single building may be guided by one person but the making of an architecture is a collective endeavor.

EUPALINOS, or THE ARCHITECT

Paul Valéry

Edited by Jill Stoner

Paul Valéry wrote the dialogue "Eupalinos" in 1922, as an introduction to a French edition of architectural plans and engravings. His interest in architecture dates to his earliest writings, most notably the "Notes on the Method of Leonardo" and the allegorical drama "Amphion." His journals attest to a critical crossroads in his own life: the choice between the arts of building and writing; and this forms the substance of the dialogue.

The speakers are Socrates, transported from antiquity to the afterworld of the present, and Phaedrus, a bedfellow from the Platonic dialogues. In absentia, *but nevertheless a catalyst for the discussion, is the architect Eupalinos, whose remembered commentaries on the subject of construction provokes Socrates to reexamine the foundations of his own philosophical enterprise. As the dialogue unfolds, Socrates accepts the role of the constructor himself, although without corporeal form through which to exercise his desires they remain in the ephemeral realm of possibility.*

Valéry remarks in a letter to a friend that the form of the dialogue was a logical extension of his own thought process: "I found myself speaking in two voices, and began to write accordingly." He notes also that the dialogue possesses the virtue of elasticity, a remark that I have reapplied to this edited version, which represents approximately half of the original and includes those exchanges that speak most directly to the subject of building. Deletions are indicated by the symbol "...................... " The full text in English translation, from which this version is taken, is in Paul Valéry, The Collected Works in English, *vol. 4 (Princeton, 1956), translated by William McCausland Stewart. It is reprinted by permission of Princeton University Press.*

A critical commentary follows. (J. S.)

Phaedrus

What are you doing here, Socrates? I have been seeking you this long time past. I have made my way through our dim habitation, I have asked for you everywhere. Everyone here knows you, and yet no one had seen you. Why did you withdraw from the other shades, and what thought has collected your soul apart from ours, on the frontiers of this transparent realm?

Socrates

Wait. I cannot answer. You know well that reflection among the dead is indivisible. We are now too much simplified not to undergo to its very end the motion of some idea. The living have a body which permits them to go out from knowledge and to enter into it again. They are like a house and a bee.

Phaedrus

Wonderful Socrates, I hold my peace.

Socrates

I thank you for your silence. By observing it, you make to the gods and to my thought the hardest of sacrifices. You have consumed your curiosity, and offered up your impatience to my soul. Speak freely now, and if there remains in you some desire to question me, I am ready to reply, now that I have finished with questioning myself and replying to myself.—But it is rare that a question that we have repressed is not instantly swallowed up.

Phaedrus

Then why this exile? What are you doing, separated from us all? Alcibiades, Zeno, Menexenus, Lysis, all our friends are surprised not to see you. They speak aimlessly, and their shades drone on.

Socrates

Look and hear.

Phaedrus

I hear nothing. I see very little.

Socrates

Perhaps you are not sufficiently dead. Here is the limit of our domain. Before you flows a river.

Phaedrus

Alas! Poor Ilissus!

Socrates

This river is the river of Time. It casts only the souls upon this bank; but it carries away everything else without effort.

Phaedrus

I begin to see something. But I can make nothing out. For an instant my gaze follows all that passes and drifts by, which it then loses without having divided it. . . . If I were not dead, this movement would nauseate me—so sad is it and irresistible. Or else I should be constrained to imitate it, as human bodies do; I should fall asleep in order to flow away myself.

Socrates

Yet this great flux is made of all such things as you have known or might have known. This vast irregular sheet of water, which rushes by without respite, rolls all colors towards nothingness. See how dim it all is.

Phaedrus

Every instant I imagine that I am going to discern some form, but what I think I have seen never succeeds in awakening the least image in my mind.

Socrates

That is because you are witnessing the true flow of beings, motionless yourself in death. We see, from this pure bank, all human things and natural forms impelled in accordance with the true speed of their essence: We are like the dreamer, in whose breast, shapes and thoughts being strangely altered by their flight, things and their transformations intermingle and are blent. Here everything is negligible, yet everything counts. Crimes engender immense benefits, and the greatest virtues develop fatal consequences: our judgment settles on nothing, idea becomes sensation before our very eyes, and every man drags after him a chain of monsters inextricably wrought of his acts and the successive forms of his body. I think of the presence and of the habits of mortals in this so fluid stream, and reflect that I was one among them, striving to see all things just as I see them at this very moment. I then placed Wisdom in the eternal station which now is ours. But from here all is unrecognizable. Truth is before us, and we no longer understand anything at all.

Phaedrus

But whence, O Socrates, can proceed that bent for the eternal which is sometimes to be noticed among the living? You pursued knowledge. Men of the grosser sort try desperately to preserve the very bodies of the dead. Others build temples

"What are you doing here, Socrates?"

and tombs which they strive to render indestructible. The wisest and best inspired of men wish to give to their thoughts a harmony and a cadence which shall guard them against change and oblivion.

Socrates

You see clearly, O Phaedrus, what folly it is! But the Fates have decreed that among the things indispensable to the race of men, there must figure some insensate desires. There would be no men without love. Nor would science exist without absurd ambitions. And whence, think you, have we drawn the primal idea and the energy for those immense efforts which have raised so many illustrious cities and useless monuments which reason admires, though she would have been incapable of conceiving them?

Phaedrus

But reason had yet some part in the matter. All, without her, would be level with the earth.

Socrates

All.

Phaedrus

Do you remember those constructions that we saw building at the Piraeus?

Socrates

Yes.

Phaedrus

Do you recall those engines, those efforts, tempered by the music of flutes; those operations so exact, that progress at once so mysterious and so clear? What confusion at the first, melting, as its seemed, into order! What solidity, what rigor were engendered between those plumb lines, and along those frail cords stretched so as to be just breasted by the growing layers of brick!

Socrates

I preserve that fine memory. O materials! Beautiful stones! . . . O how light are we become!

Phaedrus

And that temple outside the walls, near the alter of Boreas—do you remember?

Socrates

The temple of Artemis the Huntress?

Phaedrus

The very one. One day we were out there. We were discoursing of Beauty. . . .

Socrates

Alas!

Phaedrus

I was a close friend of him who built that temple. He came from Megara and was called Eupalinos. He gladly spoke to me of his art, of all the care and the knowledge that it requires; he made me understand everything that I saw with him in his workshop. I observed, above all, his astounding mind. I found in him the power of Orpheus. He foretold their monumental future to the shapeless heaps of stones and beams that lay around us; and those materials, at his voice, seemed dedicated to the one and only place to which the fates propitious to the goddess would have assigned them. How

"Here is the limit of our domain."

wonderful were his instructions to the workmen! They retained no trace of his difficult meditations of the night before. He gave them nothing but commands and figures.

Socrates

That is the very way of God.

Phaedrus

His instructions and their acts fitted together so happily that it seemed as though these men were nothing more nor less than his limbs. You cannot believe, Socrates, what a joy it was for my soul to have knowledge of a thing so well regulated. I no longer separate the idea of a temple from that of its edification. When I see one, I see an admirable action, yet more glorious than a victory and more contrary to wretched nature. Destroying and constructing are equal in importance, and we must have souls for the one and the other; but constructing is the dearer to my mind. O most happy Eupalinos!

Socrates

What enthusiasm of a shade for a phantom!—I did not know this Eupalinos. Was he, then, a great man? I perceive that he rose to the supreme knowledge of his art. Is he here?

Phaedrus

Doubtless he is amongst us; but I have never yet met him in this land.

Socrates

I know not what he could build here. Here the very projects are memories. But reduced as we are to the sole pleasures of conversation, I should be glad to listen to him.

Phaedrus

I have retained some of his precepts. I scarcely know if they would please you. They certainly enchant me.

Socrates

Can you repeat me any one of them?

Phaedrus

Well, listen. He used very often to say: "There are no details in execution."

Socrates

I understand and do not understand. I understand something, and I am not sure that it is really what he meant.

Phaedrus

Eupalinos was the man of his precept. He neglected nothing. He ordered small laths to be cut following the grain of the wood, so that, when placed between the masonry and the beams which rested thereon, they should prevent the damp from rising into the fiber and rotting it when once absorbed. He gave a like care to all the sensitive points of the building. You would have thought that it was his own body he was tending. During the process of construction he scarcely left the works. I truly believe that he knew every stone in the place. He saw to the precision of their cutting; he minutely studied all the means that have been thought of for preventing the edges from being injured and the exactness of the joining from being impaired. He directed carvings to be contrived, toothings to be left, sloping edges to be made in the marble facings. He took the most exquisite pains with the coatings of polish which he ordered to be spread over the walls of plain stone.

But all these delicate devices making for the permanence of the edifice were as nothing to those which he employed when he elaborated the emotions and vibrations of the soul of the future beholder of his work.

For the light he prepared a matchless instrument, which would redistribute it, endowing it with intelligible forms and almost musical properties, into the space where mortals move. Like those orators and poets you had in mind just now, he knew, O Socrates, the mysterious virtue of imperceptible modulations. None perceived, when confronted by a mass so delicately lightened and so simple of aspect, that he was being led to a sort of bliss by insensible curves, by minute and all-powerful inflections; and by those deep-wrought combinations of the regular and the irregular which he had introduced and concealed, and made as imperious as they were indefinable. They caused the evershifting spectator, obedient to their invisible presence, to pass from vision to vision, and from great spells of silence to mutterings of delight, according as he advanced, retreated, approached again, and as he moved within the radius of the work, impelled by its influence, and the plaything of admiration alone. "My temple," this man from Megara would say, "must move men as they are moved by their beloved."

Socrates

That is divine. I once heard, dear Phaedrus, an expression quite similar, and quite the contrary. One of our friends, whom it is useless to name, said of our Alcibiades who was so beautifully made: "Looking at him, one feels oneself becoming an architect! . . ." How I pity you, dear Phaedrus! You are much more unhappy here than I. I cared for the True alone; I gave my life to it. Well, in these Elysian meadows, though I still wonder if I have not made a rather sorry bargain, I can still imagine that there remains something for me to know. I gladly seek among the shades, the shade of some truth. But you, whose desires were formed and whose acts were governed by Beauty alone, are left resourceless here.

. . . .

What is most beautiful finds no place in the eternal!

"... the mysterious virtue of imperceptible modulations."

Socrates

Where, then, do you place it?

Phaedrus

Nothing beautiful is separable from life, and life is that which dies.

Socrates

So we may say.... But most people have of Beauty an immortal notion of some kind.

Phaedrus

I tell you, Socrates, that Beauty, according to the Phaedrus I was ...

Socrates

Is Plato not in these parts?

Phaedrus

I am speaking against him.

Socrates

Well, speak!

Phaedrus

... does not reside in certain rare objects, nor even in those models situate outside nature, and contemplated by the noblest souls as the examplars of their designs and the secret types to which their works would conform; sacred things, which should be spoken of in the very words of the poet: Gloire du long désir, Idées!

Socrates

What poet?

Phaedrus

The very admirable Stephanos, who appeared so many centuries after us. But to my mind, the idea of these Ideas, of which our wonderful Plato is the father, is infinitely too simple, and as it were too pure, to explain the diversity of Beauties, the change of preferences in men, the effacement of so many works that were cried up to the skies, the completely new creations, and the resurrections which it is impossible to foresee. And there are many other objections!

Socrates

But what is your own thought?

. . . .

It is that which puts him, without effort, above his own nature.

Socrates

"Without effort?" "Above his nature?"

Phaedrus

Yes.

Socrates

"Without effort?" How is that possible? "Above his nature?" What does this mean? I inevitably picture a man trying to climb up on his own shoulders! ... Repelled by this absurd image, I ask, Phaedrus, how can one cease to be oneself, and then return to one's essence? And how can this take place without violence?

I well know that the extremes of love, and excess of wine, or again the astounding effect of those vapors that the Pythia breathes in, take us, as the saying goes, out of ourselves; and I know even better, by my very certain experience, that our souls can, in the very heart of time, make for themselves sanctuaries impenetrable to duration, eternal in their inner selves, but transient with regard to nature; where they at last are what they know; where they desire what they are; where they feel themselves to be created by what they love, and render back to it light for light, and silence for silence, giving themselves and receiving themselves again without borrowing aught from the stuff the world is made of, nor from the Hours. They are then like those sparkling calms, circumscribed by tempests, which shift from place to place on the seas. What are we during these abysses? They imply the life they suspend. . . .

But these marvels, these spells of contemplation, these ecstasies do not illuminate for me our strange problem of beauty. I am unable to connect these supreme states of the soul with the presence of a body or of some object which brings them into being.

Phaedrus

O Socrates, it is because you always wish to draw everything out of yourself! . . . You, whom of all men I admire, you, more beautiful in your life, more beautiful in your death, than the most beautiful of visible things; great Socrates, adorable ugliness, all-powerful Thought that changest poison into an elixir of immortality, O thou who, grown cold, and the half of thy body already marble, the other still vocal, didst speak to us lovingly in the language of a god, let me tell thee what thing was perhaps wanting in thy experience.

Socrates

It is somewhat late to acquaint me with it now. But speak all the same.

Phaedrus

One thing, Socrates, one single thing you lacked. You were a divine man, and you had perhaps no need of the material beauties of the world. You scarce tasted of them. You did not, I know, disdain the sweetness of the countryside, the splendor of the city, nor yet the freshness of spring water, nor the delicate shade of the plane tree; but they were for you merely the distant ornaments of your meditations, the delightful environment of your doubtings, a favorable site for your inward steps. And as what was most beautiful would lead you far away from itself, you were always seeing something else.

Socrates

Man, and the mind of man.

Phaedrus

But still, did you never meet among men some whose singular passion for forms and appearances surprised you?

Socrates

No doubt.

Phaedrus

And yet whose intelligence and virtues were inferior to none?

Socrates

To be sure I did!

Phaedrus

Did you place them higher or lower than the philosophers?

Socrates

That depends.

Phaedrus

Did their object seem to you more or less worthy of search and of love than your own?

Socrates

It is not a question of their object. I cannot think that there exists more than one Sovereign Good. But what remains obscure to me and difficult to understand is that men so pure, as regards their intelligence, should have need of sensible forms and bodily graces to attain their highest state.

Phaedrus

One day, dear Socrates, I spoke of this very thing with my friend Eupalinos.

"Phaedrus," he was saying to me, "the more I meditate on my art, the more I practice it; the more I think and act, the more I suffer and rejoice as an architect and the more I feel my own being with an ever surer delight and clarity.

"I lose myself in long spells of expectation; I find myself again by the surprises I give myself; by means of the successive steps of my silence, I advance in my own edification; and I approach to such an exact correspondence between my aims and my powers, that I seem to myself to have made of the existence that was given me a sort of human handiwork.

"By dint of constructing," he put it with a smile, "I truly believe that I have constructed myself."

Socrates

To construct onself, to know oneself—are these two distinct acts or not?

Phaedrus

. . . and he added: "I have sought accuracy in my thoughts,

so that, being clearly engendered by the consideration of things, they might be changed as though of their own accord into the acts of my art. I have apportioned my attentions; I have arranged the problems in another order; I begin where I finished off formerly, so as to go a little further. . . . I am niggardly of musings, I conceive as though I were executing. No more now, in the shapeless void of my soul, do I contemplate those imaginary edifices, which are to real edifices what chimeras and gorgons are to true animals. But what I think, is feasible, and what I do, is related to the intelligible. . . . And then. . . . Listen, Phaedrus," he went on to say, "that little temple, which I built for Hermes, a few steps from here, if you could know what it means to me!—There where the passer-by sees but an elegant chapel—'tis but a trifle: four columns, a very simple style—there I have enshrined the memory of a bright day in my life. O sweet metamorphosis! This delicate temple, none knows it, is the mathematical image of a girl of Corinth, whom I happily loved. It reproduces faithfully the proportions that were peculiarly hers. It lives for me! It gives me back what I have given it. . . ."

. . . .

Eupalinos went on to give me a magnificent picture of those gigantic constructions which we admire in ports. They advance into the sea. Their arms, of a hard, absolute whiteness, circumscribe the lulled docks whose calm they defend. They keep them in security, peacefully gorged with galleys, sheltered by the bristling breakwaters and the resounding piers. High towers, where someone keeps watch, where during impenetrable nights the flame of pine cones dances and rages, command the offing at the foaming extremities of the moles. . . . To venture upon such works is to brave Neptune himself. You must fling mountains by the cartload into the waters you wish to enclose. You must pit the coarse rubble drawn from the depths of the earth against the shifting depths of the sea, and against the monotonous onslaught of the riders, urged on by the wind as it sweeps by. . . . Those ports, those vast ports, said my friend, what a sudden brightness in the mind! How they develop their several parts! How they descend to their task!—But the wonders peculiar to the sea, and the accidental statuary of the shores are given to the architect by a grace of the gods. All conspires towards the effect produced on our souls by these noble, half-natural constructions: the presence of the pure horizon, the waxing and the waning of a sail, the emotion that comes of being severed from the earth, the beginning of perils, the sparkling threshold of lands unknown; and the very eagerness of the men, ready as it is to change into a superstitious dread the moment they give in to it and set foot aboard. . . . Admirable theaters they are, in truth; but let us place above them the edifices of art alone! It is necessary to abstract oneself from the spells of life and from immediate enjoyment, even if for this purpose we must make a stern effort against ourselves. What is most beautiful is of necessity tyrannical. . . .

. . . .

O Phaedrus, when I design a dwelling (whether it be for the gods, or for a man), and when I lovingly seek its form, studying to create an object that shall delight the view, that shall hold converse with the mind, that shall accord with reason and the numerous proprieties . . . I confess, how strange soever it may appear to you, that it seems to me my body is playing its part in the game. . . . *Let me explain. This body is an admirable instrument, of which I am sure that those who are alive and who all have it at their disposal do not make full use. They draw from it only pleasure, pain, and indispensable acts, such as living. Sometimes they become identical with it; sometimes again they forget its existence for a space; and, at one moment mere brutes, at another pure spirits, they know not what multitudinous bonds with all things they have in themselves, and of what a marvelous substance they are made. And yet it is through this substance that they participate in what they see and what they touch: they are stones, they are trees; they exchange contacts and breaths with the matter that englobes them. They touch, they are touched; they have, and lift, weight; they move, and*

". . . those gigantic constructions which we admire in ports."

carry their virtues and vices about; and when they fall into a reverie or into indefinite sleep, they reproduce the nature of waters, they turn into sand and clouds. . . . On other occasions they store up thunderbolts and hurl them abroad! . . .

. . . .

Socrates

All this rings strange in the place where we are. Now that we are deprived of our bodies, we must assuredly bewail it, and consider that life which we have left with the same envious eye with which we formerly looked on the garden of happy shades. . . . Neither our works nor our desires follow us here; but there is room for regrets.

Phaedrus

These groves are haunted by shades eternally miserable. . . .

Socrates

If I met this Eupalinos, I should ask him something further.

Phaedrus

He must feel himself the most accursed of the blest. What would you ask him?

Socrates

To explain somewhat more clearly about those buildings of which he said "that they sing."

Phaedrus

I see that phrase of his haunts you.

Socrates

There are sayings that are bees for the mind. They have the persistence of those insects that harass us. This one stung me.

Phaedrus

And what says the sting?

Socrates

It ceases not to spur me on to expatiate upon the arts. I compare them, I distinguish them; I would hear the song of the columns and visualize in the pure sky the monument of a melody. This conceit very easily leads me to set on one side Music and Architecture; and over against them, the other arts. A painting, dear Phaedrus, covers a mere surface such as a panel or a wall: and thereupon it feigns objects and personages. Statuary likewise never adorns more than a portion of our view. But a temple, along with its precincts, or again the interior of this temple, forms for us a sort of complete greatness within which we live. . . . We are, we move, we live inside the work of man! There is not a part of that triple extent that has not been studied out and reflected upon. In it we breathe in, as it were, the will and preferences of an individual. We are caught and mastered within the proportions he has chosen. We cannot escape him.

Phaedrus

No doubt.

Socrates

But do you not see that the same thing happens to us in another circumstance?

Phaedrus

What thing?

Socrates

Being inside a work of man as fishes are in the sea, being entirely immersed in it, living in it, and belonging to it?

Phaedrus

I cannot guess.

Socrates

Why, did you never experience this when present at some religious festival, or when taking part in some banquet, while the instruments filled the hall with sounds and phantoms? Did it not seem to you that an intelligible and changing space was substituted for primitive space; or, rather, that time itself surrounded you on all sides? Did you not live in a mobile edifice, incessantly renewed and reconstructed within itself, and entirely dedicated to the transformations of a soul none other than the soul of extension itself? Was there not a changing plenitude, analogous to a continuous flame, illumining and warming your whole being by an unceasing fiery consumption of memories, of forebodings, regrets, and premonitions, and of an infinity of emotions having no clear cause? And did not those moments, and their ornaments, and those dances without dancers, and those statues, bodiless and featureless (and yet so delicately outlined), seem to surround you, slave as you were of the general presence of Music? And that inexhaustible production of enchantments, were you not enclosed along with it, nay forcibly locked up, like a Pythia in her chamber of vapors?

Phaedrus

Yes, certainly. And I even have observed that to be in that enclosure and in that universe created by sounds, wherever it be, was to be outside of oneself. . . .

Socrates

And further! Have you not felt this mobility as motionless compared to your still more mobile thought? Have you not considered, at certain instants, and as though apart from yourself, that edifice of apparitions, of transitions, of conflicts, and of indefinable events, as a thing from which one can divert one's attention, and to which one can return, as by a road, to find it all nearly the same?

Phaedrus

I confess that it has happened to me to detach myself unwittingly from music, and in some sort to leave it where it was. . . . My attention strays from it, the music itself being the starting point that invites my distraction. Then I return into its bosom.

Socrates

All that mobility forms then, as it were, a solid. It seems to exist in itself, like a temple built about your soul; you can go out and away from it; you can go in again by another door. . . .

Phaedrus

That is so. And I should even say that one never goes in again by the same door.

Socrates

There are then two arts which enclose man in man; or, rather, which enclose the being in its work, and the soul in its acts and in the productions of its acts, as our former body was entirely enclosed in the creations of its eye, and surrounded with sight. By means of two arts it wraps itself up, in two different ways, in their inner laws and wills, which are figured forth in one material or another, stone or air.

Phaedrus

I well see that Music and Architecture have each of them this profound kinship with us.

Socrates

Both occupy the totality of one sense. We only escape from the one by an inner severance; from the other only by movements. And each of them fills our knowledge and our space with artificial truths, and with objects essentially human.

Phaedrus

Therefore these two arts, being so directly related to us, without intermediaries, must also have with one another a peculiarly simple relationship?

". . . the song of the columns. . ."

Socrates

Quite so; and you rightly say: "Without intermediaries." For the visible objects on which the other arts and poetry draw—flowers, trees, living beings (and even the immortals)—when they are employed by the artist, do not cease to be what they are, nor to mix their nature and their own significance with the design of him who uses them to express his will. Thus the painter who desires that a certain place on his picture be green, there puts a tree; and by so doing he says something more than what he wished to say initially. He adds to his work all the ideas that spring from the idea of a tree, and cannot confine himself to what suffices. He cannot separate color from some being.

Phaedrus

Such is the profit, and such the disadvantage of being enslaved to real objects; each of them is a plurality of things for man and may be turned to a plurality of different uses for his acts. . . . What you say of the painter makes me think, too, of children whom a pedagogue requires to reason upon Achilles and the tortoise, and to find out how much time a hero will need to overtake a sluggish animal. Instead of dispelling the fable from their minds, and of concentrating on the numbers and their mathematical relations, they conjure up on the one hand the winged feet, on the other the tardy tortoise; their minds enter successively into the two beings; they think the one and think the other; and creating thus two incommunicable times and spaces, never reach the state where there is no more Achilles and no more tortoise, nor even time nor speed, but numbers alone and numerical relations.

Socrates

But the arts of which we speak should, on the contrary, by means of numbers and relations of numbers, engender in us not a fable, but that hidden power which makes all fables. These arts raise the soul to the creative pitch, and make it sonorous and fertile. The soul responds to that pure, material harmony which they communicate to her, by an inex-

haustible abundance of explanations and myths which she engenders without effort; and she creates, for that invincible emotion which calculated forms and right intervals impose on her, an infinitude of imaginary causes, which make her live a thousand marvelously sudden lives that merge into one another.

. . . .

Socrates

Yet these marvels are but the supreme effects of language.

Phaedrus

And so language is a constructor? . . . I already knew that it was the source of fables; and for some, even the father of the . . .

Socrates

Phaedrus, Phaedrus, impiety lacks grace in these abodes. Since there are here no thunderbolts, there is no merit in blaspheming. . . . And these dim meadows grow no hemlock. But most certainly the word can build, just as it can create, as it can corrupt. . . . An altar raised to the word should present three fronts diversely adorned; and if I had to figure forth the word in human semblance, I should give it three faces: the one, almost shapeless, would signify the common word—that which is no sooner born than it dies, and is there and then swallowed up by its very use. Immediately it is transformed into the bread that we ask for, the road we are shown, the anger of him who suffers an insult. . . . But the second face would fling from its rounded mouth a crystalline flood of eternal water: it would have the noblest features, a grand and impassioned eye; a neck powerful and massive, such as statuaries give to the Muses.

Phaedrus

And the third?

Socrates

By Apollo, how am I to figure it? . . . It would have to be some inhuman countenance, with features severe and subtle as those which the Egyptians, it is said, were able to give to the faces of their gods.

Phaedrus

And truly said. Craft, deep enigmas, an almost cruel precision, an implacable and half-bestial cunning, all the signs of feline watchfulness and of a fierce spirituality are visible in the images of those stern deities. The skillfully proportioned blend of acuteness and coldness produces in the soul a peculiar sense of uneasiness and disquietude. And these monsters of silence and lucidity, infinitely calm, infinitely alert, rigid and seemingly endowed with imminence, or with a suppleness about to be, have the semblance of Intelligence herself, in guise of beast and animal—impenetrable—all-penetrating.

Socrates

What is there more mysterious than clarity? . . . What more capricious than the way in which light and shade are distributed over hours and over men? Certain peoples lose themselves in their thoughts; but for us Greeks all things are forms. We retain only their relations; and enclosed as it were in the limpid day, Orpheuslike we build, by means of the word, temples of wisdom and science that may suffice for all reasonable creatures. This great art requires of us an admirably exact language. The very word that signifies language is also the name, with us, for reason and calculation; a single word says these three things. For what is reason if not discourse itself, when the meanings of the terms are properly limited and assured of their permanence, and when these immutable meanings are fitted to one another, and combined clearly? And that is one and the same thing as calculation.

. . . .

Phaedrus

You seem to have been won over yourself to the adoration of architecture! Here you are, unable to speak without borrowing from the major art its images and its firm ideal.

Socrates

I am still steeped in the sayings of Eupalinos that you were recalling. They have aroused within me something that is akin to them.

Phaedrus

So there was an architect in you?

Socrates

Nothing can beguile, nothing attract us, nothing makes us prick up our ears, or holds our gaze, nothing by us is chosen from among the multitude of things, and causes a stir in our souls, that was not in some sort pre-existent in our being or secretly awaited by our nature. All that we become, even fleetingly, was prepared beforehand. There was within me an architect whom circumstances did not fashion forth.

Phaedrus

How do you know this?

Socrates

By I know not what deep intent to build, which darkly troubles my thoughts.

Phaedrus

You showed no sign of it when we existed.

Socrates

I told you that I was born several *and that I died* one. *The child when it appears is a countless crowd, which life reduces soon enough to a single individual, the one who manifests himself and who dies. A multitude of Socrateses were born with me, from whom little by little the Socrates stood out who was destined for the magistrates and the hemlock.*

Phaedrus

And what has become of the others?

Socrates

Ideas. They have remained in the condition of ideas. They came, asking to be, and they were refused. I kept them within myself, as my doubts and contradictions. . . . Sometimes these germs of persons are favored by circumstance, and then we are on the verge of changing our natures. We find in ourselves tastes and gifts which we did not suspect: the musician becomes a general, the pilot feels himself to be a physician; and he whose virtue was wont to admire and respect itself discovers within himself a hidden Cacus, and the soul of a thief.

Phaedrus

It is very true that certain periods in man's life are like the crossing of ways.

Socrates

Adolescence is more particularly placed in the midst of divergent roads. . . . One day, one of my glorious days, dear Phaedrus, I experienced a strange hesitation between my souls. Chance played in my hands the most ambiguous object imaginable. And the infinite reflections that it caused me to make were equally capable of leading me to that philosopher that I became, and to the artist that I have never been. . . .

Phaedrus

It was an object that solicited you so variously?

Socrates

Yes. A paltry object, just something I found as I was walking. It was the origin of a thought divided, of itself, between constructing and knowing.

Phaedrus

Marvelous object! Object comparable to that box of Pandora, in which all good things and all evils were enclosed together! . . . Make me see this object, as the great Homer makes us admire the shield of the son of Peleus!

Socrates

It is impossible to describe. . . . Its importance is inseparable from the embarrassment which it caused me.

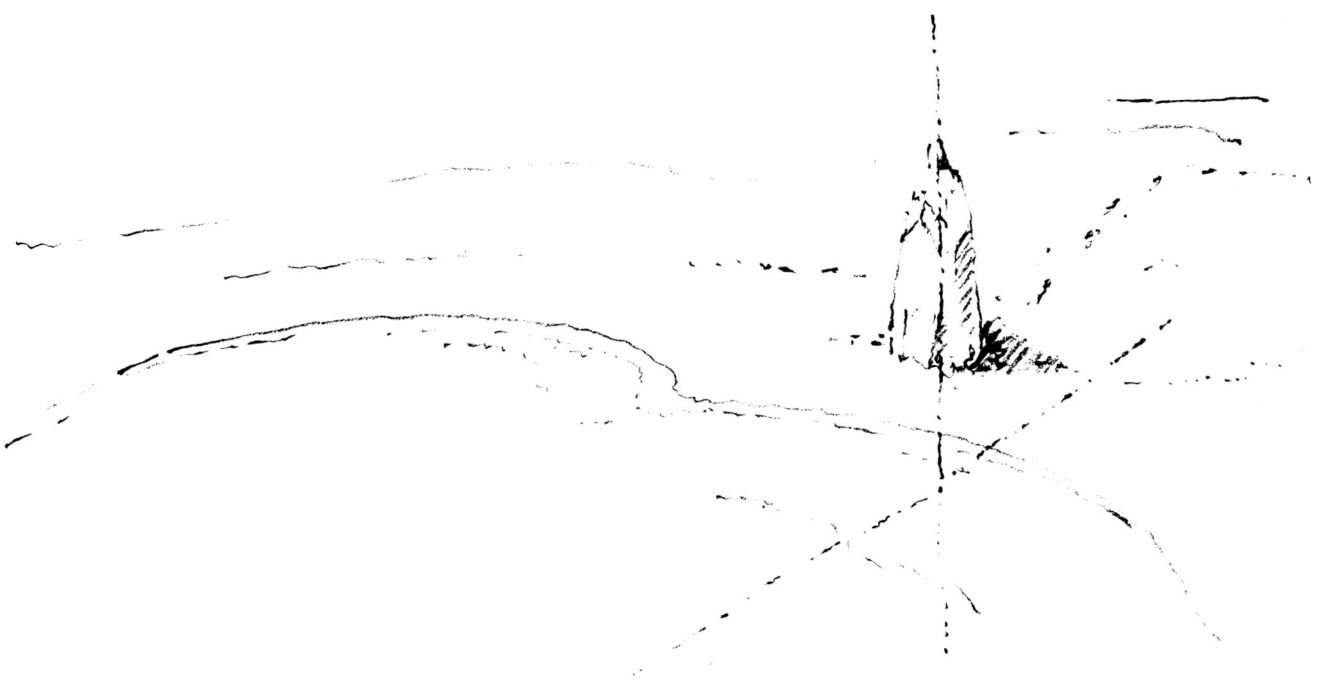

"It was the origin of a thought divided. . ."

"I was following an endless shore. . ."

Phaedrus

Explain yourself more abundantly.

Socrates

Well then, Phaedrus, this is how it was. I was walking on the very edge of the sea. I was following an endless shore. . . . This is not a dream I am telling you. I was going I know not whither, overflowing with life, half-intoxicated by my youth. The air, deliciously rude and pure, pressing against my face and limbs, confronted me—an impalpable hero that I must vanquish in order to advance. And this resistance, ever overcome, made of me, too, at every step an imaginary hero, victorious over the wind, and rich in energies that were ever reborn, ever equal to the power of the invisible adversary. . . . That is just what youth is. I trod firmly the winding beach, beaten and hardened by the waves. All things around me were simple and pure: the sky, the sand, the water. I watched, as they came from the offing, those mighty shapes which seem to be running from the coasts of Libya, charioting their glistening summits, their hollow valleys, their relentless energy from Africa all the way to Attica across the immense liquid expanse. At last they come upon their obstacle, the very plinth of Hellas; they shatter themselves against those submarine foundations; they recoil in disorder towards the origin of their motion. When the waves are thus destroyed and confounded, yet seized in turn by those that follow them, it is as though the forms of the deep were engaged in strife. The countless drops break their chains, a sparkling spray goes up. One sees white horsemen leaping beyond themselves, and all those envoys of the inexhaustible sea perishing and reappearing, with a monotonous tumult, on a gentle, almost imperceptible slope, which all their vehemence, though it be come from the most remote horizon, will yet never be able to surmount. . . . Here the foam, flung farthest by the highest wave, forms yellowish iridescent heaps which burst in the sunlight, or which the wind sweeps along and disperses in the drollest fashion, like beasts scared by the sudden bound of the sea. But as for me, I was reveling in the newborn virgin foam. . . . Its contact is of a strange softness. It is a milk—warm, airy—that comes in with a voluptuous violence, pours round one's bare feet, bathes them, passes beyond, and flows down upon them again, moaning with a voice that forsakes the shore and withdraws into itself; while the human statue, a living presence, sinks a little deeper into the sand that draws it down; and whilst the soul gives itself up to that so puissant and so delicate music, is soothed by it, and follows it eternally.

Phaedrus

You make me live again. O language charged with salt! O words truly born of the sea!

Socrates

I have let my talk run on. . . . We have all eternity during which to discourse upon time. We are here—Danaïdlike—to empty out our spirits.

Phaedrus

But the object?

Socrates

The object lay upon the beach where I was walking, where I have halted, where I have spoken to you at length of a sight with which you are as familiar as I, but which, when recalled in this place, acquires a kind of novelty from the fact that it has forever disappeared. So wait, and in a few words I shall have found this thing that I was not looking for.

Phaedrus

We are still on the seashore?

Socrates

Necessarily. This frontier between Neptune and Earth, ever disputed by those rival divinities, is the scene of the most dismal and most incessant commerce. That which the sea rejects, that which the land cannot retain, the enigmatic bits of drift;

the hideous limbs of dislocated ships, black as charcoal, and looking as though charred by the salt waters; carrion horribly pecked and washed sleek by the waves; elastic weeds torn by the tempests from the transparent pasture grounds of Proteus' flocks; collapsed monsters, of cold, deathly hues; all the things, in short, that fortune delivers over to the fury of the shore, and to the fruitless litigation between wave and beach, are there carried to and fro; raised, lowered; seized, lost, seized again according to the hour and the day; sad witnesses to the indifference of the fates, ignoble treasures, playthings of an interchange as perpetual as it is stationary. . . .

Phaedrus

And it was there that you made your find?

Socrates

Yes, there. I found one of those things cast up by the sea; a white thing of the most pure whiteness; polished and hard and smooth and light. It shone in the sun on the licked sand, that is somber and spark-bestrewn. I took it up; I blew upon it; I rubbed it against my cloak, and its singular shape suspended all my other thoughts. Who made thee? I pondered. Thou resemblest nothing, and yet thou art not shapeless. Art thou a sport of nature, O nameless thing, that art come to me by the will of the gods, in the midst of the refuse that the sea this night has flung from her?

Phaedrus

How large was this object?

Socrates

About as big as my fist.

Phaedrus

And made of what matter?

Socrates

Of the same matter as its form: matter for doubt. It was perhaps some fishbone weirdly worn by the rubbing of the fine sand under the water? Or a piece of ivory carved for I know not what purpose, by some craftsman beyond the seas? Who knows? . . . A divinity, perhaps, that perished with the very vessel which it had been made to preserve from shipwreck? But who can have been the maker of this? Was it a mortal obedient to an idea, who, pursuing with his own hands an aim foreign to the substance he is attacking, scratches, cuts off, or joins together again; pauses and considers; and finally takes leave of his work—something telling him that the work is finished? . . . Or was it not rather the product of a living body, which, unwittingly, labors with its own substance, and blindly forms for itself its organs and its armor, its shell, its bones, its weapons; causing its nourishment, which it gathers from around it, to participate in the mysterious construction which ensures it a certain lastingness?

But perhaps it was but the fruit of an infinite lapse of time. . . . Through the eternal laboring of the sea waves, a fragment of rock, by dint of being rolled and knocked to and fro, if its substance is of an unequal hardness and so is not liable in time to become rounded, may very well take on some striking semblance. It is not altogether impossible that, if a quite shapeless piece of marble or stone be given up to the permanent commotion of the waters, it should some day be thence recovered, by a chance of another kind, and should then bear the likeness of Apollo. I mean that the fisherman who had some notion of those divine features, might recognize them in this piece of marble drawn from the waters; but as for the thing itself, the sacred face is for it one fleeting form out of the family of forms that the action of the seas must impose upon it. Centuries cost nothing, and he who disposes of them changes what he will into what he will.

Phaedrus

But then, dear Socrates, when an artist makes immediately, and by a consistent application of his will, such a bust (as that of Apollo), is not his work in some sort the contrary of indefinite time?

Socrates

Precisely. It is the exact contrary, as though acts illuminated by a thought abridged the course of nature; and so we may safely say that an artist is worth a thousand centuries, or a hundred thousand, or even many more!—In other words, that this almost inconceivable stretch of time would have been necessary for ignorance or change blindly to bring to pass the selfsame thing that our skilled man has accomplished in a few days. Here is a strange standard for measuring works!

Phaedrus

Strange indeed. It is a great misfortune that we can hardly make use of it. . . . But tell me, what did you do with that thing in your hand?

Socrates

I stood still for some little time, examining it on all sides. I questioned it without stopping at an answer. . . . I could not determine whether this singular object were the work of life, or of art, or rather of time—and so a freak of nature. . . . Then suddenly I flung it back into the sea.

Phaedrus

The water splashed up, and you had a feeling of relief.

Socrates

The mind does not reject a riddle so easily.

. . . .

Intrigued by this object the nature of which I could not get to know, and which was equally claimed and rejected by all the categories, I sought to escape from the perplexing image of my find. How was this to be managed, unless by proceeding indirectly and magnifying the difficulty itself? After all, I said to myself, the same embarrassment which this object I have found causes me can be conceived about an object that is known. But with this latter, since it is known, we are in possession of both the question and the answer; or rather we are more particularly in possession of the answer, and, feeling that we have it, we omit to ask the question. . . . Suppose now that I am looking at some very familiar thing such as a house, a table, a wine jar; and that I pretend for a brief space that I am a complete savage who has never seen such objects, I might well doubt whether those objects were of human fabrication. . . . Not knowing to what use they can be put, nor even if they are of use to anyone, and being moreover informed by nobody, I should needs have to imagine some means for setting my mind at rest concerning them. . . .

Phaedrus

And what did you imagine?

Socrates

Seeking, finding, losing, and rediscovering the means of discriminating between what is produced by nature and what is made by men, I remained standing for some time in the same place, my eye hesitating amid various lights.

. . . .

I reasoned thus: a tree, with its leaves, is a product of nature. It is an edifice of which the parts are the leaves, the branches, the trunk, and the roots. I assume that each of its parts gives me the idea of a certain complexity. I then say that the whole of this tree is more complex than any one of its parts.

. . . .

I forthwith inferred that, in all these beings, the degree or level of the whole is necessarily to be ranged higher than the level of the details or rather that it can be equal to or higher than the latter, but never inferior to it.

. . . .

But with the objects which are the work of man it is quite different. Their structure is . . . a disorder!

. . . .

The acts of the man who builds or fashions something are not concerned with "all" the qualities of the substance that those acts modify, but only with some. That which is sufficient for our purpose is what matters to us. It is the effects of language that suffice the orator; its relations and its sequence that suffice the logician; and just as the one neglects rigor, the other neglects ornament. And likewise, in the material order: a wheel, a door, a tub require each a certain solidity, a certain weight, a certain adaptability in their adjustment and use; and if chestnut, elm, and oak are equally or almost equally suitable for this, the wheelwright or carpenter will use them more or less indifferently with an eye only to the expense. But in nature you do not see the lemon tree bearing apples, although in a given year they would perhaps cost her less than lemons to produce.

Man, I assert, fabricates by abstraction, ignoring and forgetting a great part of the qualities of what he uses, and concerning himself solely with clear and definite conditions, which can most often be simultaneously satisfied, not by a single material, but by several kinds. He drinks milk or wine or water or ale indifferently out of gold, glass, horn, or onyx; and whether the vase be wide or slender, or shaped like a leaf or a flower, or with a quaintly twisted foot, the drinker considers the drinking alone. Even he who made that cup was unable to do more than harmonize very roughly its substance, its form, and its function. For the intimate subordination of these three things and their profound interfusion could be but the work of Creative Nature herself. The artisan cannot do his work without violating or disarranging an order by the forces which he applies to matter in order to adapt it to the idea he wishes to imitate, and to the usage he intends.

. . . .

Phaedrus

I see that, impelled by that object found on the seashore, to which none but you would have paid the least attention, your adolescent genius rose almost at once to the consideration of a very important and very simple distinction. From the slightest of incidents you extracted this thought: that human creations resolve themselves into a conflict between two kinds of order, one of which, being natural and given, suffers and sustains the other, which is the expression of the needs and desires of man.

Socrates

That is what I believed. Man does not need the whole of nature, but only a part. The philosopher is one who conceives a wider view, and wishes to have need of everything. But the man who only wishes to live has need neither of iron nor of bronze "in themselves"; but merely of such and such a hardness or ductility. He is forced to take them where they are to be found, that is to say in a metal which has also other qualities indifferent to him. . . . He considers his aim alone. If he wishes to drive in a nail, he strikes it with a stone, or with a hammer made of iron, or of bronze, or even of very hard wood; and he drives it in with little blows, or with a single more forcible one, or sometimes by a pressure; what does it matter to him? The result is the same, the nail is driven home. But if one is not concerned to follow the thread of this action, and if one simply views all the circumstances, these operations appear to be each entirely different—phenomena that cannot be compared with one another.

Phaedrus

I now understand how you could hesitate between constructing and knowing.

Socrates

One has to choose between being a man and being a mind.

Man can act only because he can ignore, and content himself with a part of this knowledge which is his peculiar extravagance, a knowledge that is somewhat more extensive than is necessary!

Phaedrus

Yet it is this slight excess that makes us men!

Socrates

Men? . . . Think you that dogs do not see the stars, for which they have no use? It would be enough for them that their eye should perceive terrestrial things; but it is not so exactly adapted to pure utility as to be blind to the celestial bodies and the majestic ordering of night.

Phaedrus

They never tire of howling at the moon!

Socrates

And do not humans strive in a thousand ways to fill or to break the eternal silence of those infinite spaces that affright them?

Phaedrus

Your own life was spent so! . . . But for my part, I cannot console myself for the death of that architect who was in you, and whom you assassinated by meditating overmuch on the fragment of a shell! With your profundity and prodigious subtleties, Socrates, you should have far surpassed our most famous builders. Neither Ictinos, nor Eupalinos or Megara, nor Chersiphron of Gnossos, nor Spintharos of Corinth would have been able to rival Socrates the Athenian.

Socrates

Phaedrus, I beg of you! . . . This subtle matter of which we are now made does not permit of our laughing. I feel I ought to laugh, but I cannot. . . . So refrain!

Phaedrus

But, quite seriously, Socrates, what would you have done, as an architect?

Socrates

How can I tell? . . . I only see in a general way how I should have conducted my thoughts.

Phaedrus

Conduct them at least to the threshold of the edifice that you did not build.

Socrates

I have merely to pursue that sort of reverie of reasoning that I have but now been engaged in.

We have said—or as good as said—that all visible things proceed from three modes of generation or production, which moreover mingle and interpenetrate. . . . The one kind chiefly make chance manifest, as can be seen from the fragments of a rock, or from some landscape, not specially chosen, peopled with plants that have sprung up at random. The other kind—like the plant itself, or the animal, or the piece of salt, whose purple-tinted facets cohere mysteriously—lead us to imagine a growth that is simultaneous, sure, and blind, and encompassed within a duration that seems potentially to contain them. It is as though what they will be is waiting on what they were; and, further, as though they increase in harmony with their surroundings. . . . And finally there are the works of man, which, in some sort, cut across this nature and this chance, utilizing them, but doing them violence, and at the same time violated by them in accordance with what we said a moment ago.

Now the tree does not construct its branches and leaves; nor the cock his beak and his feathers. But the tree and all its parts, and the cock and all his, are constructed by the principles themselves, which do not exist apart from the constructing. That which makes and that which is made are indivisible; and it is so with all bodies that live, or that have a sort of life, like crystals. It is not acts that engender them; and their generation cannot be explained by any combination of acts, for acts presuppose living beings.

Nor yet can it be said that they are spontaneous—this word is simply an avowal of impotence. . . .

We know, moreover, that these beings have need of a thousand things in their environment, in order that they should be. They are dependent upon everything, though the action of all things seems, by itself, incapable of creating them.

But as for the objects made by man, they are due to the acts of a thought.

The principles are separate from the construction, and are as it were imposed by a tyrant from without upon the material, to which he imparts them by acts. Nature in her work does not distinguish the details from the whole; but pushes from all sides at once, chaining herself to herself, without experiments, without regressions, with no models, no special aim, and no reserves; she does not separate a project from its execution; she never takes a direct course regardless of obstacles, but compromises with them, mixes them with her motion, goes round them or makes use of them; as though the path she takes, the object that follows this path, the time spent in covering it, the very difficulties it presents, were all of the same substance. If a man waves his arm, we distinguish this arm from his gesture, and we conceive between gesture and arm a purely possible *relation. But from the point of view of nature, this gesture of the arm and the arm itself cannot be separated. . . .*

Phaedrus

So, to construct would be to create by separate principles?

"Man's deepest glances are those that go out to the void."

Socrates

Yes, the peculiarity of man is to create in two kinds of time, one of which runs on in the domain of pure possibility, in the very heart of that subtle substance which can imitate all things and combine them with one another to infinity. The other time is nature's. It in one sense contains the first, and in another sense, is contained in it. Our acts partake of both. The project is separate from the act, and the act from the result.

Phaedrus

But how can we conceive the separateness, and how find the principles?

Socrates

They are not always so distinct as I have said. All men, moreover, do not distinguish them equally. But a very simple and very primitive reflection will suffice to give an idea of it. Man discerns three great things in the All: he finds there his body, he finds there his soul—and then there is the rest of the world. Between these things there is an unceasing commerce, and sometimes even a confusion arises; but always after a certain time has elapsed, these three things come to be clearly distinguished from one another. It is as though their intermixture cannot last, and that this division must necessarily manifest itself from time to time.

Phaedrus

A man who is asleep sometimes takes his leg for a stone, and his state of rest for a movement. He takes his desire for a light; the noise of his blood for a mysterious voice; the feeling of his own face being brushed by a fly becomes the vision of a terrifying face which pursues him. . . . But all that cannot last. He wakes up, flinging the past far from his body, reserving it for his soul; anew he divides all things and rebuilds himself according to his principles.

Socrates

It is therefore reasonable to think that the creations of man are made either with a view to his body, and that is the principle we call utility, *or with a view to his soul, and that is what he seeks under the name of* beauty. *But, further, since he who constructs or creates has to deal with the rest of the world and with the movement of nature, which both tend perpetually to dissolve, corrupt, or upset what he makes, he must recognize and seek to communicate to his works a third principle, that expresses the resistance he wishes them to offer to their destiny, which is, to perish. So he seeks* solidity *or* lastingness.

Phaedrus

Yes, those are indeed the main characteristics of a complete work.

Socrates

Architecture *alone demands them, and carries them to their highest point.*

. . . .

Phaedrus, my pale Phaedrus, Brother Shade of my Shade, how infinite, I feel, would be my regrets if they had some substance to work on, and if flesh were not wanting for their exercise! They begin to stir, but it all comes to nothing! They are sketched out, but they cannot take on color! . . . Is there anything vainer than the shadow of a sage?

Phaedrus

A sage himself.

Socrates

Alas yes! a sage himself, who leaves behind him only the reputation of a talker, and various sayings abandoned to immortality. . . . What did I do but lead the rest of mortals to believe that I knew much more than they about the most doubtful things?—And the secret for getting it believed lies in a death so well conducted, adorned with such injustice, and surrounded with such friendships, that it darkens the sun, and disconcerts nature. What is more to be feared than the making of death a sort of masterpiece? . . . Life cannot defend herself against those undying death scenes. She invincibly imagines—in her simplicity—that the finest part of the tragedy begins after the last word of the last line! . . . Man's deepest glances are those that go out to the void. They converge beyond the All.

Alas! alas! I have made use of a truth and a sincerity much more deceitful than are myths and inspired sayings. I taught what I invented. . . . I got children upon the souls I seduced, and I delivered them skillfully.

Phaedrus

You are hard on all of us.

Socrates

If you had not listened to me, my pride would have sought some other way of subjugating your thoughts. . . . I should have built, sung. . . . O thoughtful waste of my days! What an artist I have destroyed! . . . What things have I scorned, and alas, what begotten! . . . I feel like the Judge of my own spiritual Netherworld pronouncing sentence against myself! Whilst the facility of my renowned sayings pursues and afflicts me, here am I raising up as Eumenides my actions that did not take place, my works that were never born—those glaring voids, vague and enormous crimes that they were: those murders, the victims of which are imperishable things! . . .

Phaedrus

Be comforted. . . . You would regret them much more if you had begotten them!

. . . .

You must not wish to begin again. . . . One never succeeds a second time. . . .

Socrates

Do not be yet more bitter.

Phaedrus

I must confess that your words have wounded my friendship a little. Surely you understand that if you lower yourself, and if you depreciate Socrates, then Phaedrus who gave himself up to him so piously, Phaedrus sees himself reduced to the last extreme of folly and blind simplicity!

Socrates

Alas! such is our condition! But I am trying to make something of it. Think you not that we ought now to employ this boundless leisure which death leaves us, in judging and rejudging ourselves unwearyingly, revising, correcting, attempting other answers to the events that took place, seeking, in fine, to defend ourselves by illusions against nonexistence, as the living do against their existence?

Phaedrus

And what is it you want to depict against this background of nothingness?

Socrates

The Anti-Socrates.

Phaedrus

I can imagine more than one. There are several contraries to Socrates.

Socrates

This one will be . . . the constructor.

Phaedrus

Good. The Anti-Phaedrus is listening to him.

Socrates

O coeternal with me in death, faultless friend, and diamond of sincerity, hear then:

It served no purpose, I fear, to seek this God, whom I have tried all my life to discover, by pursuing him through the realm of thought alone; by demanding him of that most variable and most ignoble sense of the just and the unjust, and by urging him to surrender to the solicitings of the most refined dialectic. The God that one so finds is but a word born of words, and returns to the word. For the reply we make to ourselves is assuredly never anything other than the question itself; and every question put by the mind to the mind is only, and can only be, a piece of simplicity. But on the contrary, it is in acts, and in the combination of acts, that we ought to find the most immediate feeling of the presence of the divine, and the best use for that part of our strength that is unnecessary for living, and seems to be reserved for the pursuits of an indefinable object that infinitely transcends us.

If, then, the universe is the effect of some act; that act itself, the effect of a Being, and of a need, a thought, a knowledge, and a power which belong to that Being; it is then only by an act that you can rejoin the grand design, and undertake the imitation of that which has made all things. And that is to put oneself in the most natural way in the very place of the God.

Now, of all acts the most complete is that of constructing. A work demands love, meditation, obedience to your finest thought, the invention of laws by your soul, and many other things that it draws miraculously from your own self, which did not suspect that it possessed them. This work proceeds from the most intimate center of your existence, and yet it is distinct from yourself. If it were endowed with thought, it would divine your existence, which yet it would never succeed in probing, nor in conceiving clearly. You would be for it a God. . . .

Let us now consider this great act of constructing. Note, Phaedrus, that when the Demiurge set about making the world, he grappled with the confusion of Chaos. All formlessness spread before him. Nor could he find a single handful of matter, in all this waste, that was not infinitely impure and composed of an infinity of substances.

He valiently came to grips with this frightful mixture of dry and wet, of hard and soft, of light and gloom, that made up this chaos, whose disorder penetrated into its smallest parts. He disentangled that faintly luminous mud, of which not a single particle was pure, and wherein all energies were diluted, so that the past and the future, accident and substance, the lasting and the fleeting, propinquity and remoteness, motion and rest, the light and the heavy were as completely mingled as wine with water, when poured into one cup. Our men of science are always trying to bring their minds close to this state. . . . But the great Shaper acted in contrary wise. He was the enemy of similitudes and of those hidden identities that we delight to come upon. He organized inequality. Setting his hand to the rough matter of the world, he sorted out its atoms. He divided the hot from the cold, and the evening from the morning; he drove back almost all the fire into the subterranean hollows, and hung clusters of ice on the very trellises of dawn, beneath the vaultings of the eternal Ether. By him extension was distinguished from movement, night from day; and in his rage to disunite everything, he clove asunder the first animals, which he had just disassociated from the plants, into male and female. After he had finally disentangled even that which was most mixed up in the original confusion—matter and mind—he hoisted to the loftiest empyrean, to the inaccessible peak of History, those mysterious masses, whose silent irresistible descent into the uttermost depth of the abyss begets and measures Time. He squeezed out from the mud the sparkling seas and pure waters, lifting the mountains out of the waves, and portioning out in fair islands whatever concreteness remained. Thus he made all things; and, from a remnant of mud, humankind.

But the constructor whom I am now bringing to the fore finds before him, as his chaos and primitive matter, precisely that world order which the Demiurge wrung from the disorder of the beginning. Nature is formed, and the elements are separated; but something enjoins him to consider this work as unfinished, and as requiring to be rehandled and set in motion again for the more especial satisfaction of man. He takes as the starting point of his act, the very point where the god had left off.—In the beginning, he says to himself, there was what is: the mountains and forests; the deposits and veins; red clay, yellow sand, and the white stone which will give us lime. There were also the muscular arms of men, and the massive strength of buffaloes and oxen. But there were in addition the coffers and storerooms of intelligent tyrants and of citizens grown overrich by trade. And lastly there were priests who wished to house their god; and kings so puissant that they had nothing more to desire than a matchless tomb; and republics that dreamed of inexpugnable walls; and refined archons, who had such partiality for actors and fair musicians that they were all afire to build for them, out of the state treasuries, the most resonant theaters.

Now the gods must not remain without a roof, nor souls without drama. The masses of marble should not remain lifeless within the earth, constituting a solid night; nor the cedars and cypresses rest content to come to their end by flame or by rot, when they can be changed into fragrant beams and dazzling furniture. But still less should the gold of rich men lazily sleep its heavy sleep in the urns and gloom of treasuries. This so weighty metal, when it becomes the associate of a fancy, assumes the most active virtues of the mind. It has her restless nature. Its essence is to vanish. It changes into all things, without being itself changed. It raises blocks of stone, pierces mountains, diverts rivers, opens the gates of fortresses and the most secret hearts; it enchains men; it dresses, it undresses women with an almost miraculous promptitude. It is truly the most abstract agent that exists, next to thought. But thought exchanges and envelops images only, whereas gold incites and promotes the transmutations of all real things into one another; itself remaining incorruptible, and passing untainted through all hands.

Gold, limbs, projects, various substances, all are present; yet nothing results.

Here I am, says the Constructor, I am the act. You are the matter, you are the force, you are the desire; but you are separate. An unknown industry has isolated and prepared you according to its means. The Demiurge was pursuing his own designs, which do not concern his creatures. The converse of this must come to pass. He was not concerned about the troubles that were bound to spring from that very separation

"He was the enemy of similitudes."

which he diverted or perhaps bored himself with making. He has given you the means of living, and even of enjoying many things, but not generally those which you particularly want.

But I come after him. I am he who conceives what you desire a trifle more exactly than you do yourselves; I shall consume your treasures with a little more consistency and genius than you consume them; and without a doubt I shall cost you very dear; but in the end everyone will have gained. I shall make mistakes sometimes, and we shall have some ruins; but one can always very profitably look upon a work that has failed as a step which brings us nearer to the most beautiful.

Phaedrus

It is fortunate for them that you are a dead architect!

Socrates

Must I be silent, Phaedrus?—So you will never know what temples, what theaters, I should have conceived in the pure Socratic style! . . . I was going to give you an idea how I should have carried out my work. I should first have set out all the problems, evolving a flawless method. Where?—For what?—For whom?—To what end?—Of what size?—And exercising an ever stricter control over my mind, at the highest point I should have realized the operation of transforming a quarry and a forest into an edifice, into splendid equilibriums! . . . And I was drawing up my plan with an eye to the purpose of the humans who pay me; taking into account localities, lights, shadows, and winds; choosing the site according to its size, its aspect, its approaches, the adjacent lands, and the true nature of the subsoil. . . .

Then out of raw materials I was going to put together my structures entirely ordained for the life and joy of the rosy race of men. . . . Objects most precious for the body, delightful to the soul, and such as Time itself must find so hard and so difficult to digest that he will be able to subdue them only by the assaults of centuries; and that only after having clothed them in a second beauty: a mellow gold upon them, a sacred majesty upon them, and, wrought by the flight of time, a charm which comes of comparisons growing up about them and of a secret tenderness that wraps them all about. . . . But you shall learn no more. You can conceive only the old Socrates, and your stubborn shade . . .

Phaedrus

Call it rather faithful, Socrates, faithful.

Socrates

In that case it must follow me; and change if I change!

Phaedrus

But do you mean to revoke in eternity all those sayings that made you immortal?

Socrates

Immortal there—relatively to mortals! . . . But here. . . . But . . . there is no here, *and all that we have been saying is as much a natural sport of the silence of these nether regions as the fantasy of some rhetorician of the other world who has used us as puppets!*

Phaedrus

It is in this that immortality rigorously consists.

"In that case it must follow me. . ."

Exeter Library, Louis I. Kahn, architect.

THE MEANING OF CONSTRUCTION; CONSTRUCTION AS MEANING: A COMMENTARY ON THE DIALOGUE "EUPALINOS"

Jill Stoner

Life said: "I am sad; therefore I weep."
Music said: "I weep; therefore I am sad." (*from Valéry's notebooks*)

WITH ARCHITECTURE, the most timeless of the arts, as their subject, Socrates and Phaedrus seem appropriately situated in the eternal afterworld of Valéry's poetic imagination. "Motionless themselves in death," they achieve an architectonic presence, larger than life and intransient. Socrates speaks of their surroundings as a "dim habitation"; yet as the conversation proceeds, the shadowy realm on the banks of the Illisus seems to take on shape, and an impression of built form is invoked. This impression, a product of both the subject matter of the dialogue and the form itself, can be understood as a scaffolding upon which Valéry builds a series of ideas not only on the nature of architecture, but on the broadest implications of aesthetic experience.

These ideas taken together are Valéry's implicit message to architects, addressing two significant questions: "Why build?" and "How to build?" The first of these questions explores the semantic dimensions of building, or the sense of meaning inherent in the resolution of an image and its essence. The second explores the syntactic dimension, or the sense of meaning as a process of establishing order and relationships. These two dimensions "frame" Valéry in an historical position which lies at the juncture of two aesthetic/philosophical periods: Symbolism and Structuralism. Symbolism, a theory of the poetic *image*, is largely concerned with the semantic dimension. Structuralism, a science of *relationships*, or *order*, is concerned with syntax. The dialogue addresses both sets of issues in the context of a seemingly casual and spontaneous conversation. Because Valéry himself was independent of any aesthetic or philosophical "movement" at the time he wrote the dialogue, he offers concise yet synthetic impressions on the nature of building, and at the same time reveals a sensitivity to the humor and irony of the human condition.

Symbolism and Semantics

Symbolism emerged as an organized framework of thought and as the major force in French poetry in the latter half of the nineteenth century. Like most "movements" in the arts, it evolved as a reaction against, and an attempt to resolve, the conflicting philosophical temperaments which preceded it: on one hand the rational, analytical mode of the encyclopedic theorists, such as Diderot, Viollet-le-Duc, and Quatremere de Quincy, and on the other the purely romantic, psychological inquiry into the self in the tradition of the English romantic poets. The former sought the resolution of tensions through rational inquiry; for the latter, the act of artistic expression became a catharsis to resolve tensions within the self. The Symbolists took the raw material of the rational world, explicit and unambiguous images, and by virtue of their composition sought to evoke a response to the essence of the thing, or perhaps the essence of some thing not yet known.

It was Mallarmé, the central voice in Symbolist expression, that proposed the search for essences, the most pure objects of meaning, as the primary task of the poet. The process of creation, then, became a force which implied its own cancellation upon the attainment of the Absolute, which he characterized as the Void. From the mention of him in the dialogue ("the most admirable Stephanos"), we conclude that Socrates and Phaedrus have been partaking of their boundless leisure for almost two millenia, and thus are speaking in Valéry's time. In fact, when Socrates first mentions Plato, Phaedrus replies: "I am speaking against him," and thus revokes the notion of the pure Idea as a basis for understanding the aesthetic process. Instead, the creative act assumes the importance of mediator between the pure concepts of mind, duration, and essence, and the respectively contingent counterparts of body, temporality, and material. These three dualities form facets of the Aristotelian concept of substance and accident, the latter being a specific and tangible attribute of the former.[1] They are, in a sense, dialectics of self, of time, and of world, the explication of which substantiates the premise that built form, as a mediation between substance and accident, affords the fullest possible expression of meaning in a given time and society.

Structuralism and Syntax

That the dialogue reflects the influence of the Symbolist poets is less surprising than that it prefigures the attitudes of Structuralism as advanced by Levi-Strauss, in the field of anthropology, and de Saussure, in linguistics, some years later. In this view of the creative act, the process of composing relationships, or the syntactical dimension of the work, replaces symbolic or metaphoric associations as the primary conduit for meaning. Levi-Strauss, in his studies of the myths of pre-literate cultures, demonstrates that it is the structure of the myth, rather than the story line, that is the prevailing attribute, and that this structure is most usually perceived as a system of correspondences. Thus, Structuralism reveals meaning through order rather than through image; it is the formal investigation of syntax.

Valéry explores this notion of process, or the discovery of correspondences, through an analysis of the similarities between architecture and music, as those unique arts which are "without intermediaries." These correspondences, or "structures," appear as the active resolution of those dialectics introduced through the semantic dimension. That is, they are both the generating condition and the perceived reality of the act of construction. The Constructor, as the one who orchestrates these resolutions, puts himself "in the very place of God." This is not to say that he assumes the power of creation, but rather the responsibility of elucidation, revealing correspondences that lie hidden within the raw matter of the world-as-is. The process of construction is both ordered and intuitive; the concept of order implies a synthesis reflecting the dialectics of self, time, and world.

This synthesis of the semantic and syntactic dimension of building, offered in the casual and fortuitious dialogue between Valéry's phantom philosophers, appears some fifty years later in the buildings of Louis Kahn. In particular, Kahn's Exeter Library reveals the dialectic of substance and accident in the three semantic contexts of self, time, and world. At the same time, the library is a composition of elements through symmetries, repetitions, and similitudes that engender meaning at the level of syntax, or structure. In the discussion which follows, a diagrammatic "text" of Exeter's complex levels of meaning parallels the written text.

In order to relate thematic issues specifically to the dialogue, a series of quotations serve to introduce each section of the argument.

"One must choose between being a man and being a mind."

Eupalinos, evoked into presence through the reminiscences of Phaedrus, saw the act of the construction as the most viable resolution of the dual nature of the self as manifest by body (man) and mind. "By dint of constructing," he remarks, "I truly believe I have constructed myself," indicating that the process of building is both personal and self-fulfilling. This classic dichotomy of body and mind (the dialectic of the self) is subsequently developed in Socrates' extended anecdote of the incident on the beach, where an ambiguous object, "the origin of a thought divided, of itself, between constructing and knowing," provoked him to make the personal choice between man and mind.

In his reflection he concludes that the object's origins and significance are of equal matter, ("matter for doubt"), and that its ambiguity reveals the fundamental difference between those objects produced by nature and those produced by man. Nature produces by a process which recognizes all influences, laws, and contingencies: a process in which choice and accommodation are never a factor. Man, on the other hand, "fabricates by abstraction," such that the act of construction involves not a sequence of inevitability but a process of intuitive selection.

In another work Valéry comments further on the interdependence of construction and abstraction: "Thus it is through abstraction that the work of art can be constructed, and this abstraction is more or less energetic, more or less easy to define, to the degree that those of its elements which are taken from reality exist in more or less complex constellations."[2] Socrates had sought this process of abstraction in thought alone, believing that an involvement with the physical contingencies of the world would compromise his search for truth. He realizes only after deprived of his body, which in life had been his "only prison," the value of this incessant commerce between the corporeal and spiritual selves. In the ensuing discussion of construction by separate principles, this theme of a balance between substance and accident is extended beyond the self to the broader framework of time.

"We have all eternity in which to discourse upon time."

The dimension of time is introduced early in the dialogue as a substantive force flowing as a river, which "... casts only souls upon its bank, but carries away everything else without effort." "Everything else" refers to all that is not man, and hence that which has not the benefit of his activities. After concluding that it is man who brings the qualities of abstraction to the creative act, Phaedrus can justify the notion that "the artist's work is the contrary of indefinite time." Here again is a refutation of the Platonic notion of Beauty as that which belongs to the eternal, existing independently of both man and the changes wrought by time. He concludes: "Nothing beautiful is separable from life, and life is that which dies," meaning nothing more, perhaps, than that aesthetic expression has a double momentary aspect: the first of its creation, and the second of its perception and appreciation.

The aspect of time which contains man's creative act, specifically the act of construction (Valéry implies a construction with words as well as with stone) joins the temporal element of *process* with the concept of duration: "It is the peculiarity of man to create in two kinds of time, one of which runs in the domain of pure possibility. . . . the other of which is natures."

Thus in the decision to build, the architect separates himself from duration and, by virtue of his act, fixes himself in the present. Just as he chooses the raw materials for the work, and the resultant forms, by a process of abstraction, so does that act serve to arrest time, also by abstraction, as a moment which belongs not to nature but to man. Conversely, it is also through the act of construction that one rejoins duration; for while the artifact marks a specific time, it also addresses the principle of lastingness. Speaking of both poetry and architecture, Valéry says: "Even in the slightest of compositions, one must think of duration, that is *memory*, which is to say *form*, just as the builders of steeples and towers must think of structure."[3] Thus the concept of duration wears the double face of infinite past and infinite future, and it is the act of construction which joins the memory with lastingness in the present.

"Your being must therefore of necessity divide itself and become, at one and the same instant, hot and cold, fluid and solid, free and fast–rose, wax, and fire, matrix and Corinthian metal."

Body and mind, present and eternal, material and essence: the third of these thematic dialogues establishes the creative act as created artifact. From the Idea, or essence, one seeks physical form; from form, to distill the essence. The Symbolist poets, and Mallarmé in particular, pursued the realm of the pure Idea, and ultimately the Void, through poignant symbols of tangible objects. He described the effort in this way: "I say 'flower,' and musically there arises the very Idea of it; the one absent from all bouquets."[4] By similar argument sound engenders silence, time summons forth eternity; materials evoke negation, and

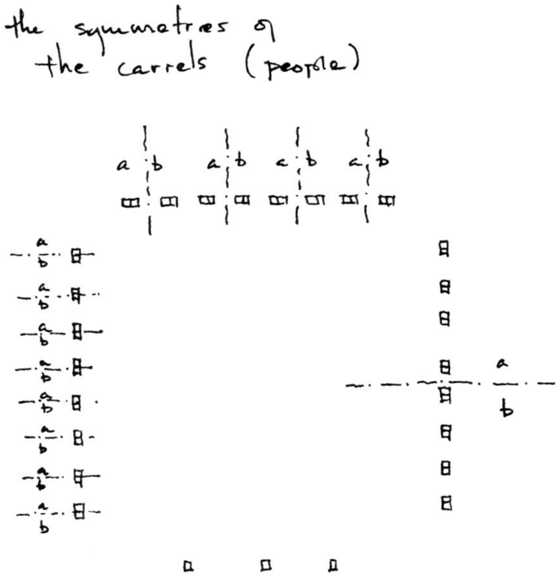

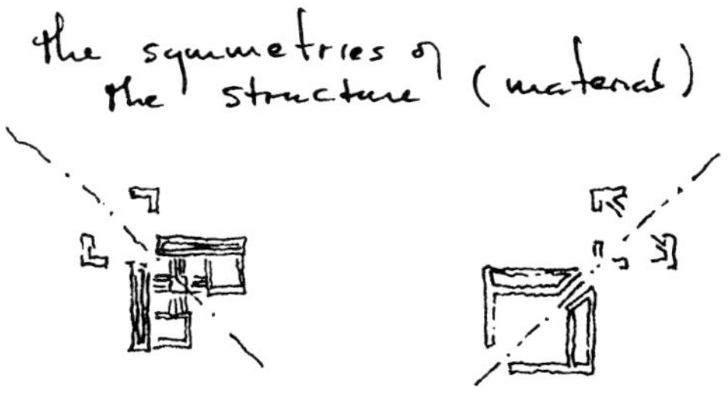

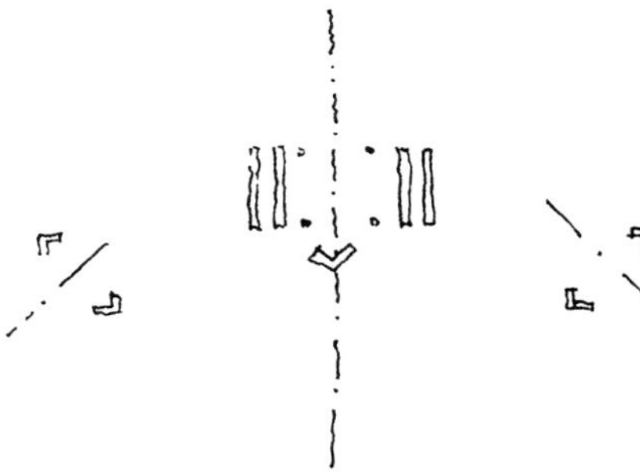

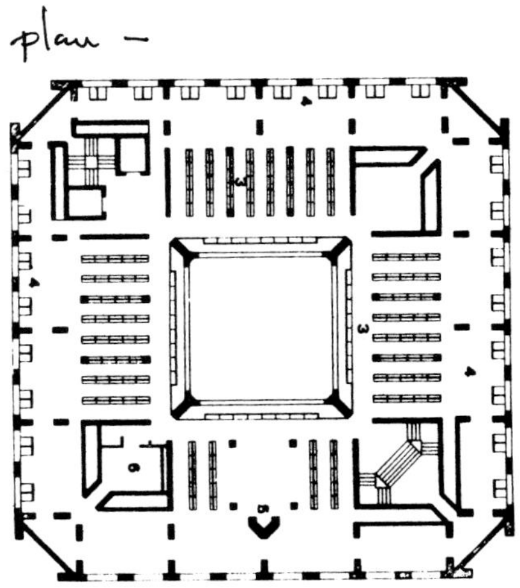

Plan, Exeter Library.

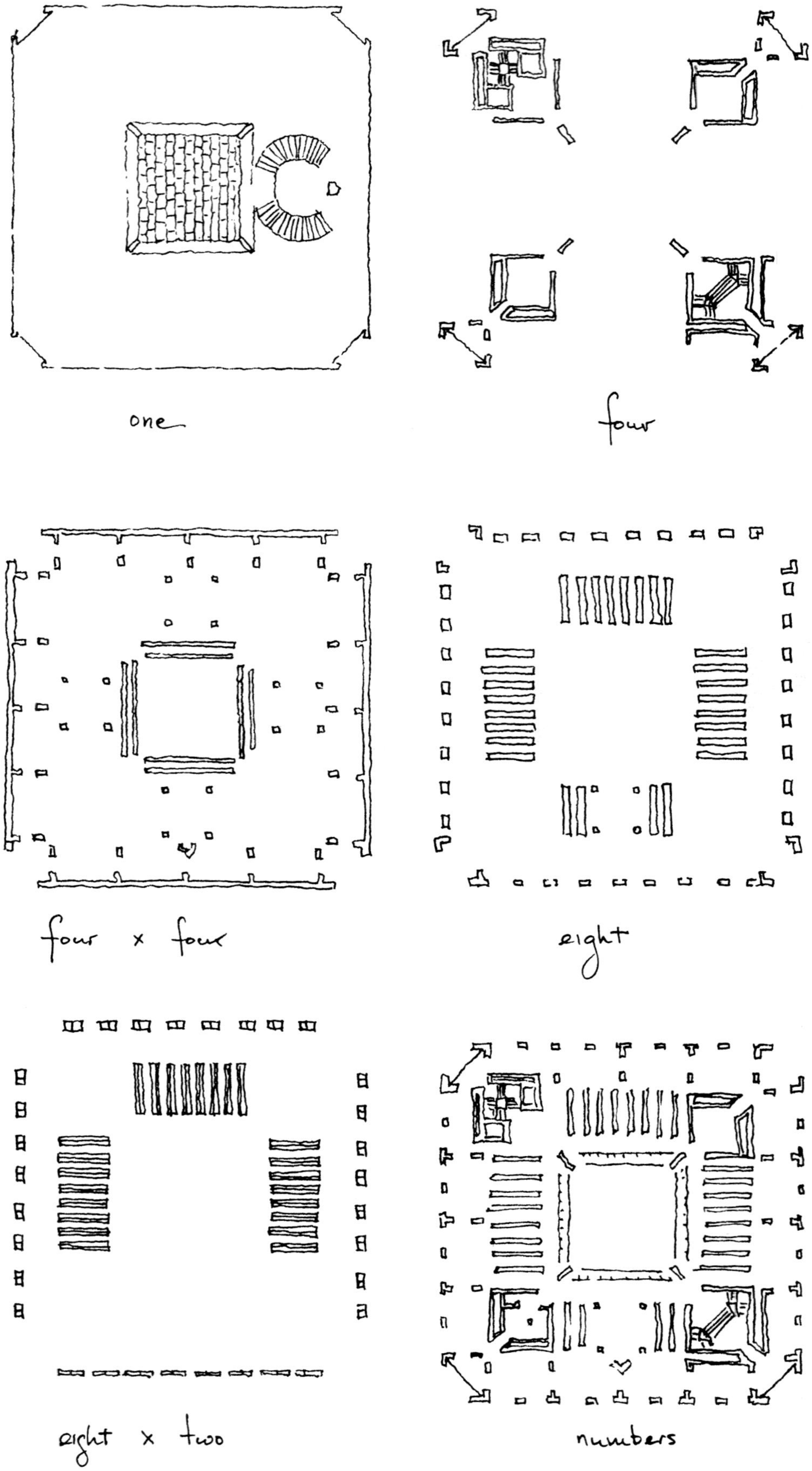

Plan, Exeter Library.

movement produces stasis. In a sense this can be characterized as a one-way aesthetic process, in which the specific image, by its juxtaposition with others in significant "constellations," dissolves into silence, darkness, the absolute.

Yet Valéry postulates a reciprocal aesthetic process through these opposing forces, as engendered through the act of construction. Speaking of the methods of Eupalinos, Phaedrus recalls that he ". . . prepared for the light a matchless instrument, which would redistribute it, endowing it with intelligible forms and almost musical properties." Conversely, his constructions became "a mass so delicately lightened" that its material properties began to dissolve into a sense of abstracted space. Thus the Idea (here represented as light) is resolved into material form, neither incidental nor inevitable, but chosen by the process of abstraction for its potential evocation and meaning.

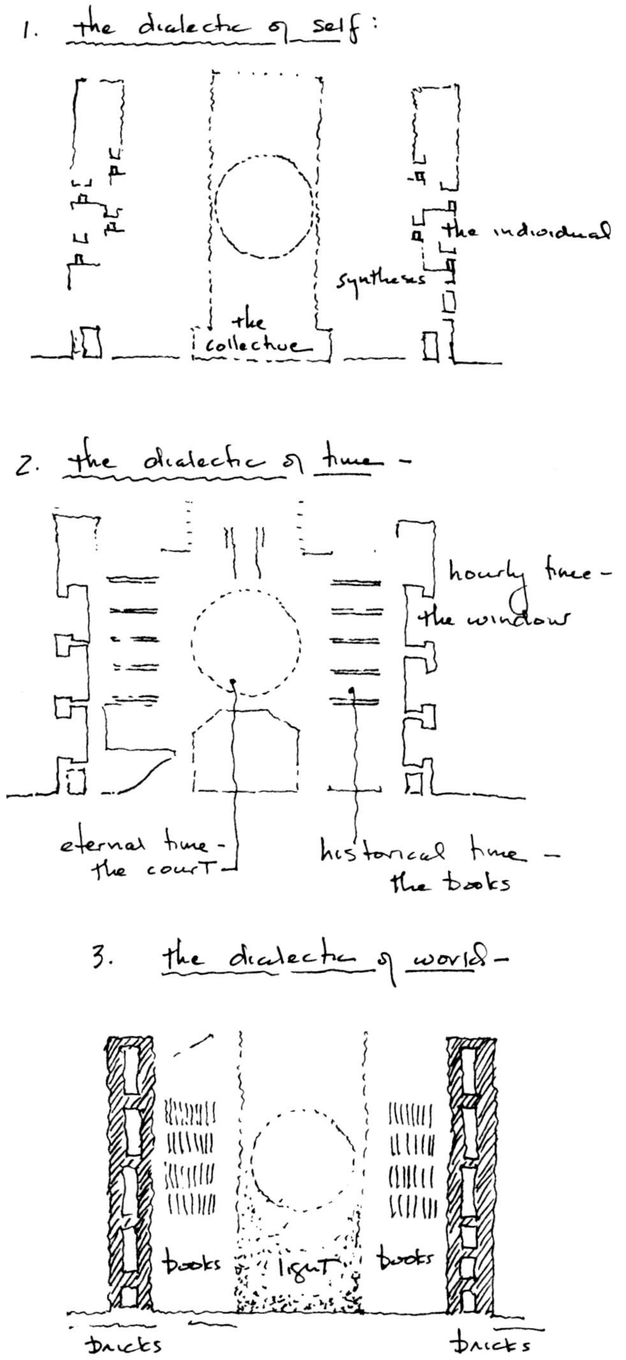

"But the arts of which we speak should, on the contrary, by means of numbers and relations of numbers, engender in us not a fable, but that hidden power which makes all fables."

Although architecture is presented in the dialogue as the primary symbol of meaning in the integration of self and world, Valéry moves beyond the condition of symbolic form in architecture to the notion of building as an expression of significant structures. This thought is developed through an extended discussion of the parallels between architecture and music. The ideas in many ways reflect the organized developments in the philosophical movement of Structuralism, as a theory of the properties and meaning of language. It has been problematic to the field, when dealing with literary and linguistic forms, that the raw material of language is, by definition, over-laden with the understood meaning of the words.[5] One may take the meaning without attention to the transcendent structure, but it is difficult if not impossible to perceive the structure independently from the meaning. One does come close to this sense of language as pure structure in the case of primitive chanting, often accompanied by music and conducted in articulated space. Collective ritual is described by Socrates as an occasion when ". . . intelligible and changing space was substituted for primitive space; or, rather, that time itself (surrounds) you on all sides."

Yet music and architecture are set apart from the other arts, specifically poetry and painting, as those forms which "dispense with imitation"; that go beyond metaphor and even symbol to the expression of the pure relation of things. The correspondence between these two arts is the subject of Valéry's play "Amphion," in which the hero, on receiving the lyre from Apollo, begins to play, whereupon the mountainside behind him transforms in cadence with the music into temples, staircases, and terraces. In a written introduction to the drama, Valéry states:

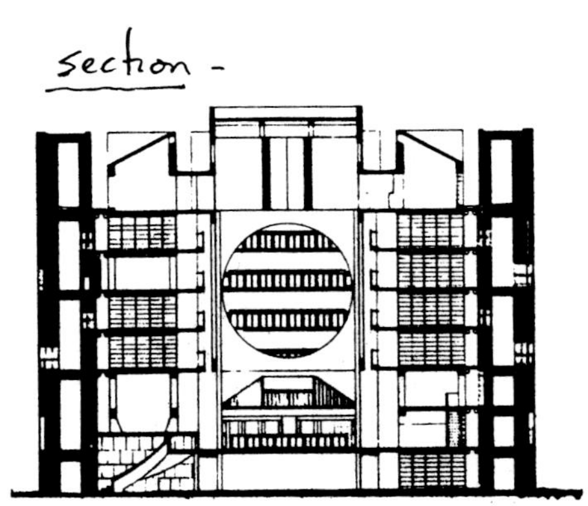

Section, Exeter Library.

> Since I thought so freely about so many things, and among them rather often about music . . . and about architecture, it was bound to happen that certain relationships between the two should come to amuse my mind. . . . It is clear that music and architecture are, both of them, arts that dispense with the imitation of objects; . . . Both admit of *repetition*, which is a very powerful instrument . . . and *composition*–which is the relation of the particular detail to the whole–is much more felt in music and architecture than in the arts whose object is the reproduction of visible things.[6]

The essence of both arts is not that to which they refer, for they refer to nothing outside of themselves. At the same time, they do not make one think of their own particular nature, but rather "enclose man in man," and also enclose man in an edifice, mobile or static, through which he moves in accordance with the structure provided. Eupalinos himself, *in absentia*, refers to the metaphoric association of architecture and music as a "divine analogy": an analogy which is reiterated throughout the dialogue in such images as a "mobile edifice," and the "song of the columns."

The powers in these arts which engender the sense of fable are the structures that transcend physical (architectonic) and temporal (musical) compositions, and attain a sense of purpose in the establishment of an order, or law. The task, Valéry tells us, is to ". . . produce essentially human objects; to make use of sensible means, which yet are not likenesses of sensible things nor duplicates of known beings, to give form and shape to laws, or deduce from the laws themselves their form and shape."[7] These laws include not only the properties of repetition and composition, but also symmetry, reflection, modulation, ornamentation, and others. Critical analysis of any work of architecture or music, and for that matter poetry as well, reveals their inherent presence. Thus, as audience and critic, we reverse the process of construction, and *deduce from the form and shape the laws themselves*. The building may then be understood as a sort of message-giver, a static and transcendent fable in stone.

Phaedrus: But are not all figures geometrical? Socrates: No more than a noise is a musical sound.

In the portion of the dialogue following the discussion of music and architecture, Socrates concludes that there is "no geometry without the word." "Word" in this sense refers not to the specific denotative elements of language, but to the most general communicative powers of man. In fact, it is a mythic property of man's being-in-the-world, signified as "Word," that has appeared often in the literature of this century as the symbol of human desire.[8] The notion of the word is then synonymous with man's unequivocal search for meaning, and it is that which is the motivating force behind the act of construction.

In the built forms of primitve societies the process of construction is largely intuitive; yet the constructions themselves are not accidental. Geometric properties appear without having been given a name. In fact the properties such as symmetry and repetition have no meaning in and of themselves, but have acquired meaning in the expression of human institutions. Each institution is itself the expression of some collective desire, however general, whose meaning is essential to our particular societies.

In the light of these forays into the formal and mythic properties of the act of construction, Socrates undergoes his significant transformation from sage to would-be constructor. In accepting the role of constructor, he assumes responsibility for the collective voice, and in a final and eloquent soliloquy sets forth a program for the art of building. His premise assumes a synthesis of the aspects of semantics and syntax as they relate to architecture. Finally, the dialogue itself assumes a symmetrical structure, a reflection between beginning and end, in both the image of a timeless chaos and the complex transformation of Socrates himself.

"Truth is before us, and we no longer understand anything at all."

At the beginning of the dialogue we are at the end of time, or in endless time, "that flux made of all such things as you have known or might have known." At the end of the dialogue, we are at the beginning of time, where the original Creator faced "all formlessness spread before him." The similarity of these two images is heightened in the comparison of Socrates-the-sage, the personality that dominates in the opening scene, and that of the Demiurge, the central figure in the account of creation. Each grappled with sorting out the totality of the world: Socrates in the world of ideas, (the semantic dimension) and the Creator in the world of material things (the syntactic dimension). In each case, reason and analysis prevailed, and the product reflected the vain pursuits of the master.

In contrast to both his corporeal self and the Demiurge is the Anti-Socrates who appears at the end: "Here I am, says the Constructor, I am the act." His method is intuitive, he seeks out the similitudes that the Great Shaper had obliterated in his systemization of all things. He agrees that "reason must play some part in the game"; yet understands that desire is the essence of the creative impulse. While Eupalinos was interested in construction as a process of self-actualization and fulfillment (he designed a small temple to reflect the proportions of the woman he loved), Socrates as Constructor assumes a collective voice. "I am he who understands what you desire a trifle more exactly than you do yourselves." Through intuition and interpretation, he proposes to build structures ordained entirely for the "rosy race of men." But needless to say his ambitions cannot be realized. Immaterial as time itself, his own insubstantial image exaggerates the supreme irony of the human condition; yet in his revised wisdom he offers a significant inspiration for the building of architecture.

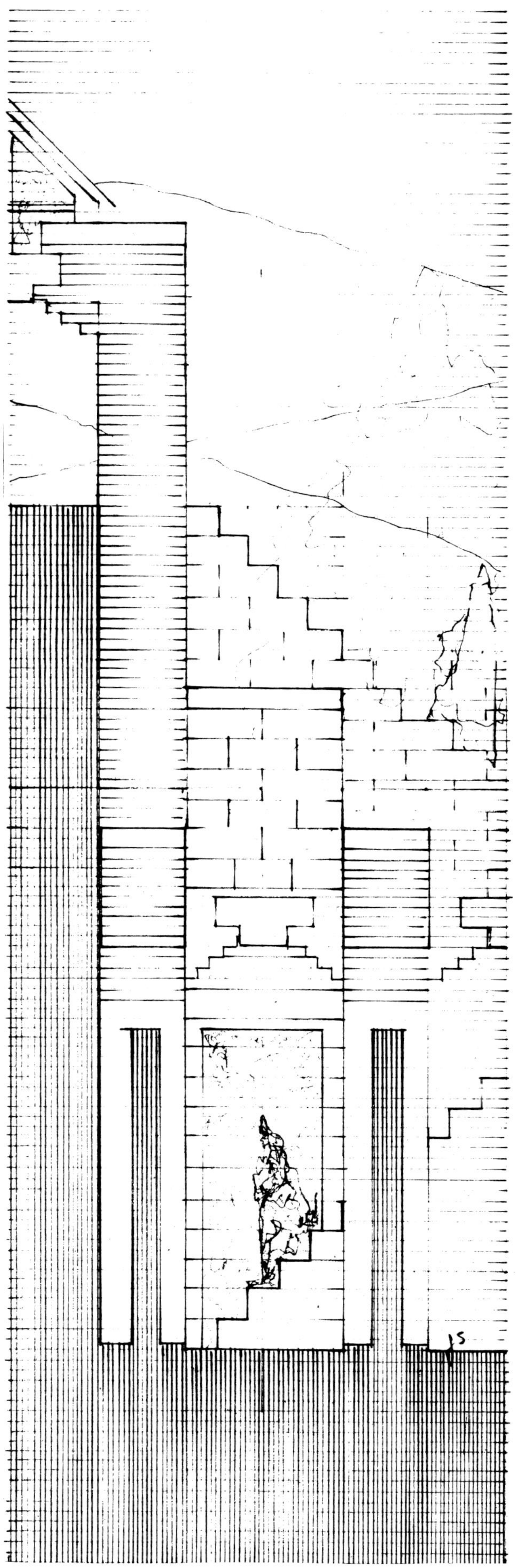

NOTES

1. Aristotle, *The Poetics*, trans. Ernest Barker, (Oxford, London, 1958).

2. Paul Valéry, *Varieté*, p. 266, trans. Ralph Freedman, in "Paul Valéry: Protean Critic," *Modern French Criticism*, ed. John K. Simon, (Chicago, 1972).

3. Paul Valéry, "Amphion," Introduction to the Audience," in *Complete Works*, Volume 3, (Princeton, 1960).

4. Stephan Mallarmé, *Divigations, Oeuvres Compiet*, (Gallimard, Paris, 1945). My translation.

5. Roland Barthes, *Elements of Semiology*, trans. (Lavers and Smith, New York, 1967), pp. 13–15.

6. Paul Valéry, "Amphion," Introduction to the Audience.

7. Paul Valéry, letter to Paul Souday, quoted in Wallace Stevens, preface to the Dialogues, *Complete Works*, Volume 4, (Princeton, 1958).

8. Word in this sense becomes synonymous with myth, or fable, as the expression of memory, desire, and value. See:
Jorge Luis Borges, *Labyrinths*, (New York, 1962).
"At the beginning of literature was the fable, and also at the end."
Dylan Thomas, "The Mouse and the Woman," in *Adventures in the Skin Trade*, (New York, 1970).
"Look at the millions of stars, he said. They make some pattern in the sky. It is a pattern of letters spelling a word. One day I shall look up and read the word."
———, "In the beginning," in *Collected Poems*, (New York, 1971).
"In the beginning was the word, the word
That form the solid bases of the light
Abstracted all the letters of the void;
And from the cloudy bases of the breath
The word flowed up, translating to the heart
First characters of birth and death.
Umberto Eco, *The Name of the Rose*, trans. William Weaver, (New York, 1983).
"In the beginning was the Word, and the Word was with God, and the Word was God."
Paul Valéry, "A Fond Note on Myth," *Complete Works*, (Princeton, 1963). Vol. 10.
"In the beginning was the fable."

BIOGRAPHIES

Emilio Ambasz is an architect, graphic, and industrial designer. He was born in Argentina in 1943 and received a Master of Architecture degree from Princeton University where he has also taught. He is the author of numerous articles and books on architecture and design and served as the Curator of Design at the Museum of Modern Art in New York from 1970–77. At present, he is the President of the Architectural League. His architectural projects have been built in South America, Europe, and the United States.

Hajj Abdullah Nuridin Durkee was born in Warwick, N.Y. in 1938. He studied as an apprentice at numerous ateliers and fine arts institutes and pursued his religious education at *Jamiat Um al-Qurra, Colleat ash-Shariat* in Makka, Saudi Arabia. Presently he is studying under the hand of Sheikh Doctor Ibrahim al-Battawi abu-Dhikri of Azhar University in Cairo, Egypt. His work has been exhibited in major museums including the Museum of Modern Art in New York and the San Francisco Museum of Fine Arts. The author of numerous articles, Mr. Durkee has founded or co-founded several organizations including the Dar al-Islam Foundation in Abiquiu, New Mexico (1981). He is professionally associated with High Altitude and Desert Research Associates (HADRA).

Marco Frascari is an Assistant Professor of Architecture and Assistant Director of the Ph.D. Program in Architecture at the University of Pennsylvania. Born in Mantua, Italy in 1945, Dr. Frascari received *Dottore in Architettura* from the *Istituto Universitario di Architettura* in Venice in 1969. He studied architectural graphics and design at Columbia and New York University and received an M.S. in Architecture from the University of Cincinnati in 1978. In 1981 he was awarded a Ph.D. in Architecture from the University of Pennsylvania. A member of the Semiotic Society of America, he has written and lectured extensively on Semiotics both in the U.S. and abroad.

Else Glahn is a Professor of Chinese Language and History at the University of Aarhus in Denmark. Once a practicing architect, Professor Glahn conducts research in classical Chinese texts, particularly those pertaining to architecture and Song Dynasty history.

Karsten Harries is a Professor of Philosophy at Yale University. Born in Jena, Germany, Professor Harries attended Yale where he received B.A. and Ph.D. degrees. His work has centered on the thought of Martin Heidegger and recently his interests have focused increasingly on the philosophy of architecture. He has written numerous articles and is the author of *The Meaning of Modern Art* (Northwestern University Press, 1968) and *The Bavarian Rococo Church: Between Faith and Aestheticism* (Yale University Press, 1983). The present essay, together with its companion pieces in *Perspecta* 17, 19, and 20 (Yale University, M.I.T. Press, 1980, 1982, 1983) is an examination of ideas which should develop into a more comprehensive study.

Herman Hertzberger, Dutch architect and teacher, was born in Amsterdam in 1932. He graduated from the University of Technology at Delft in 1958 after which he opened his own practice in Amsterdam which he currently maintains. A Professor of Architecture at Delft, Mr. Hertzberger has been a Visiting Professor at numerous universities in the U.S. and abroad including Columbia, Harvard, and Toronto Universities, M.I.T., and the University of Pennsylvania. He was an editor of (Dutch) Forum (with Van Eyck, Bakema and others) from 1959–1963. His completed projects have been featured in many international journals and magazines.

Liane Lefaivre studied psychology and comparative literature at McGill University and at the *Université de Strasbourg*. She has been working on problems of cultural history and theory. She has published, in collaboration with Alexander Tzonis, numerous articles in architectural journals and scholarly reviews. She is the author of a book, *Documentary History of Architecture (1000–1800)*, to be published in 1985.

Cesar Pelli is Dean of the School of Architecture at Yale University and principle of Cesar Pelli and Associates in New Haven, Connecticut. Born in 1926 in Argentina, Mr. Pelli received a *Diploma of Arquitecto* from the *Universidad Nacional de Tucuman* in Argentina in 1949 and an M.S. in Architecture from the University of Illinois in 1954. He worked in the office of Eero Saarinen and Associates, was Vice President for Design at Daniel, Mann, Johnson, and Mendenhall (DMJM) in Los Angeles, and was Partner for Design at Gruen Associates, also in Los Angeles. He has been a Visiting Professor of Architecture at the *Universidad Nacional de Tucuman* and UCLA and was William Henry Bishop Visiting Professor of Architecture at Yale University. His work, both with Gruen and Cesar Pelli and Associates, has been widely published both in the U.S. and abroad.

David C. S. Polk is an architect and Adjunct Associate Professor of Architecture at the University of Pennsylvania. Born in 1931 in Baltimore, Maryland, Mr. Polk is a graduate of Dartmouth College and received a Bachelor of Architecture degree from the University of Pennsylvania in 1956. From 1965–1968 he worked on several projects while employed in the office of Louis I. Kahn. He is presently in private practice in Philadelphia and a Visiting Critic in architecture at the University of Southern California.

A. T. Seymour III is president of Thos. S. Byrne Inc., General Contractors, of Fort Worth, Texas. Born in 1928, he graduated from the University of Texas at Austin in 1950 with a B.S. Degree in Architectural Engineering. Mr. Seymour is a past president of the Fort Worth Chapter of the Associated General Contractors and was named "Man of the Year" by that organization in 1978. He recently completed seven years on the Advisory Board to the School of Architecture at the University of Texas at Austin and was actively involved in the creation of the architecture chair at the University named for the noted architect O'Neil Ford. Mr. Seymour is a past president of the Fort Worth Museum of Science and History, the Exchange Club of Fort Worth, and is a member of the American Arbitration Association.

Jill Stoner was born in New London, Connecticut in 1953. She studied literature and philosophy at New College and received a Master of Architecture degree from the University of Pennsylvania in 1979. She has taught at the Pennsylvania State University and presently teaches at the Philadelphia College of Art and the University of Pennsylvania.

George Thomas is an Architectural Historian and principle member of Clio Group, Inc., an architectural restoration consulting and research firm in Philadelphia. Professor Thomas attended Dickinson College and the University of Pennsylvania where he was awarded a Ph.D. in 1975. He is a Lecturer in Art History at the University of Pennsylvania, Assistant Professor at Drexel University, and was Centennial Professor at Bryn Mawr College in 1982–83. He has written numerous articles on architectural history and urbanism and is the co-author of two books, *Cape May: Queen of the Seaside Resorts*, with Carl Doebley (Art Alliance Press: Philadelphia, 1976) and *The Architecture of Frank Furness*, principle essay by James O'Gorman (Philadelphia Museum of Art, 1973).

Alexander Tzonis is Crown Professor of Architectural Methodology at the University of Technology at Delft, the Netherlands. He studied at Yale University and has taught at Yale and the Graduate School of Design at Harvard University. The director of several research projects, he served as a consultant to the *United Nations Habitat Conference*. He has been the editor of a multi-disciplinary series on man-made environment for Penguin Books, is General Editor of the *Architectural Archives* (Garland Press, New York), and is on the editorial staff of the Dutch architectural review *Bouw*. He is the author of *Towards a Non-Oppressive Environment* (I Press, dist. by M.I.T. Press, 1972), *Shape of Community*, with Serge Chermayeff (Penguin Books, 1971), and *Classicism in Architecture: The Imperative for Order* (S.U.N., Netherlands, 1983, Blume, Spain, 1984).

ILLUSTRATION CREDITS

"Editor's Introduction"—Behrens
Photographs by Roy Vollmer

"A House Built on Sand"–Thomas
All illustrations are courtesy of the author.

"The Tell–the–Tale Detail"–Frascari
All illustrations are by the author except as noted below:
2. Paul Gerhardt, *International Competition for a New Administrative Building for the Chicago Tribune*, Chicago, 1923.

3.Eugenio Battisti, *Filippo Brunelleschi*, Rizzoli, New York, 1981.

4. G. B. Piranesi, "The Polemical Works", J. Wilton-Ely, Ed., Gregg Institute, 1972.

6. *Encyclopédie*, Diderot and D'Alambert, 1762–77, pl. 15, vol. 7.

7, 8. Franco Borsi, *Leon Battista Alberti*, Electa Editrice, 1975.

9. E. Viollet-Le-Duc, *Dictionnaire Raisonné de L'Architecture*, Paris, 1868, p. 507, vol. 9.

10. Frank Lloyd Wright, 1953, from Arthur Drexler, *The Drawings of Frank Lloyd Wright*, Bramhall House, New York, 1962. Copyright The Frank Lloyd Wright Foundation. Reproduced with permission.

11, 12, 18, 27, 28. Photographs of Scarpa drawings by author.

14, 15, 16. Foto Giacomelli.

17. Giuseppe Zambonini, in *Progressive Architecture*, May 1981. Reproduced with permission.

"Building Order"–Hertzberger
All illustrations are courtesy of the office of Herman Hertzberger except as noted.
13. Willem Diepraam.

"On Truth and Lie in Architecture"–Harries
1. *The Seven Lamps of Architecture*, John Ruskin, 1849.

2. Conway Library, Courtald Institute of Art, University of London.

3. Reproduced by permission of Holt, Rinehart and Winston, Publishers, from Peter Blake, *God's Own Junkyard*, Copyright 1964 by Peter Blake.

4. Yale University.

5. University of Pennsylvania Archives.

6. *AIA Journal*, May 1983, photograph by Harlan Hambright.

7. Ezra Stoller, ESTO Photographics.

8. Hirmer Fotoarchiv, Munich.

9. Anthony Fisher.

10. *AIA Journal*, September, 1977.

11. Office of Michael Graves.

12. Office of Michael Graves.

13. *AIA Journal*, May 1983, photograph copyright by Norman McGrath.

"Hassan Fathy in New Mexico"–Durkee
All illustrations are provided by the author and copyright, Center for Traditional Islamic Crafts except as noted below:
2. Antoine Predock.

3. Reproduced by permission of G. E. Kidder Smith. Copyright 1976.

5, 10, 11, 28, 32, 35, 36. Hassan Fathy, El Sabah International Institute for Appropriate Technology, 1980, 1981.

7. Copyright Ch. Bastin & J. Evrard. Reproduced with permission.

8, 15, 27. Copyright Edward Klamm, 1980. Reproduced with permission.

14, 17. Rick Homans, *Adobe Today*.

19. *Progressive Architecture*, June 1983, after drawing by Hassan Fathy. Reproduced with permission.

20, 21, 22, 23, 33. Chant Avidessian, Copyright, 1983, The Aga Khan Award for Architecture. Reproduced with permission.

"The Immeasurable Made Measurable: Building the Kimbell Art Museum"–Seymour.
1, 2, 5, 6, 7, 9, 10, 11, 12, 13, 14, 15, 16, 17, 18, 19 Kimbell Art Museum, Fort Worth, Texas: photographs by Bob Wharton.

3, 4, 8 Thos. S. Byrne, Inc., Fort Worth, Texas.

"Fabula Rasa"–Ambasz
E. Allen, from *Stone Shelters* (Cambridge, Mass., 1969).

"*Yingzao Fashi*: Chinese Building Standards in the Song Dynasty"—Glahn
Illustrations 2, 4, 5, 8, 9, 11, 12, 13, 14, 15, 16, 17, 18, and 19 are adapted by the author from Liang Sicheng: "Song Yingzao fashi tuzhu," Qinghua daxue jianzhu xi, Beijing, n.d.
Illustrations 20, 21 and 22 are from Qi Yingshou: "Hebei sheng Xincheng xian Kaishan si dadian." Wenwu 1957, vol. 10.
Other sources are as follows:
1. Tobias Faber: "En Kina-rejse." Copenhagen 1979.

3. Guo Daiheng: "Lun Zhongguo gudai mugou jianzhu di moshuzi." Jianzhu shi luwen ji. Vol. 5, 1981.

6. Liu Zhiping: "Zhongguo jianzhu leixing ji jiegou." Beijing 1957.

7. Shujing toshuo. 1905.

10. From the scroll: "Qingming shanghe" by Zhang Zeduan, ca. 1117.

23. Liang Sicheng: Qingdai yingzao suanli ji zeli. Beijing 1934.

24. Author.

"The Expression of Meaning and the Necessity for Integration" –Polk
Illustrations are courtesy of the author except as noted.
2. Emil Schulthess, from *Africa* (New York, 1959).

3, 4. Henri Cartier-Bresson, from *The Decisive Moment* (New York, 1952).

6. Tet Borsig, from *Designs in Nature* (New York, 1962).

7. Courtesy of the Kimbell Art Museum.

9, 10. E. Allen, from *Stone Shelters* (Cambridge, Mass., 1969).

11. Lawrence Weaver, *Houses and Gardens by Sir Edwin Lutyens, R.A.*

13. N. Ardalan and L. Bakhatiar, *A Sense of Unity*, (Chicago, 1973).

"The Mechanization of Architecture and the Birth of Functionalism"–Tzonis and Lefaivre

Photo credits may be found in the captions except as noted.
32. Courtesy of the office of James Stirling.

"Architectural Form and the Tradition of Building"–Pelli
Illustrations are courtesy of the office of Cesar Pelli and Associates except as noted.
4. Arthur Golding.

6. S. Hill.

8. Balthazar Korab.

11. Mitsuo Matsuoka, Japan Architect.

17. Wolfgang Hoyt, ESTO Photographics.

20. Balthazar Korab.

21. Kenneth Champlin.

23. Rendering by Lebbeus Woods.

"Eupalinos, or The Architect"–Valery
All drawings by Jill Stoner. Photograph by C. Duncan.

"The Meaning of Construction and the Construction of Meaning" –Stoner
All drawings are by the author. Photographs by C. Duncan.

"Biographies"
Drawing by Salih Memecan.